D0857838

Spatial Multimedia and Virtual Reality

Spatial Multimedia and Virtual Reality

EDITED BY

ANTONIO S. CAMARA
New University of Lisbon

JONATHAN RAPER
City University, London

First published 1999 by Taylor & Francis
11 New Fetter Lane, London EC4P 4EE

Simultaneously published in the USA and Canada by Taylor & Francis
325 Chestnut Street, 8th Floor, Philadelphia PA 19106

Taylor & Francis is an imprint of the Taylor & Francis Group

British Library Cataloguing in Publication Data

A catalogue record for this book is available from the British Library.
ISBN 0-7484-0819-3 (cased)
ISBN 0-7484-0820-7 (paper)

Library of Congress Cataloging in Publication Data are available

Cover design by Hybert Design and Type
Printed and bound by T.J. International Ltd, Padstow, Cornwall

Contents

II. VIRTUAL REALITY

Preface

The GISDATA Program of the European Science Foundation brought together researchers interested in the application of multimedia and virtual reality to the solution of spatial problems to a meeting in Rostock in May 1994. The Conference on Spatial Multimedia and Virtual Reality, partially sponsored by GISDATA, and held in Lisbon in October 1995, was attended by this group and several other specialists from around the world. In September 1997, at the final GISDATA Conference in Strasbourg, the views on the current and future spatial applications of multimedia and virtual reality were again presented by some of the authors included in this Monograph.

The Research Monograph in Spatial Multimedia and Virtual Reality is, thus, dedicated to present research efforts of the past three years mostly associated with the GISDATA initiative. The Monograph includes selected updated articles on visualization, virtual reality, multimedia and simulation. Most of these contributions are related to environmental applications.

In some of the Monograph articles, one can sense the already existing possibilities offered by the new digital media to the handling of geographical information (GI). There are also forward-looking proposals that are still in an alpha stage. However, all of the articles do represent non-traditional views in the GI which, we hope, will stimulate the current and the next generations of researchers in the field.

ANTÓNIO S. CÂMARA
Monte de Caparica

JONATHAN RAPER
London

Series introduction

Welcome

The *Research Monographs in Geographical Information Systems* series provides a publication outlet for research of the highest quality in GIS, which is longer than would normally be acceptable for publication in a journal. The series includes single- and multiple-author research monographs, often based upon PhD theses and the like, and special collections of thematic papers.

The need

We believe that there is a need, from the point of view of both readers (researchers and practitioners) and authors, for longer treatments of subjects related to GIS than are widely available currently. We feel that the value of much research is actually devalued by being broken up into separate articles for publication in journals. At the same time, we realise that many career decisions are based on publication records, and that peer review plays an important part in that process. Therefore a named editorial board supports the series, and advice is sought from them on all submissions.

Successful submissions will focus on a single theme of interest to the GIS community, and treat it in depth, giving full proofs, methodological procedures or code where appropriate to help the reader appreciate the utility of the work in the Monograph. No area of interest in GIS is excluded, although material should demonstrably advance thinking and understanding in spatial information science. Theoretical, technical and application-oriented approaches are all welcomed.

The medium

In the first instance, the majority of Monographs will be in the form of a traditional textbook, but, in a changing world of publishing, we actively encourage publication on CD-ROM, the placing of supporting material on web sites, or publication of programs and of data. No form of dissemination is discounted, and prospective authors are invited to suggest whatever primary form of publication and support material they think is appropriate.

The editorial board

Future submissions

Anyone who is interested in preparing a Research Monograph should contact either of the editors. Advice on how to proceed will be available from them, and is treated on a case by case basis.

For now we hope that you find this, the **sixth** in the series, a worthwhile addition to your GIS bookshelf, and that you may be inspired to submit a proposal too.

Editors:	Professor Peter Fisher	Professor Jonathan Raper
Addresses:	Department of Geography University of Leicester Leicester LE1 7RH UK	Department of Information Science City University Northampton Square London EC1V 0HB UK
Telephone:	+44 (0) 116 252 3839	+44 (0) 171 477 8415
Fax:	+44 (0) 116 252 3854	+44 (0) 171 477 8584

List of contributors

David Aragão
Environmental Systems Analysis Group
New University of Lisbon
2825 Monte de Caparica
Portugal
Telephone: (351) (1) 295 44 64
E-mail: dga@students.fct.unl.pt

Ralf Bill
Institute for Geodesy and Geoinformatics
University of Rostock
D 18051 Rostock
Germany
E-mail: bill@agr.uni-rostock.de
http://Web.agr.uni-rostock.de/iggi/iggi.html

Lars Bodum
GISplan
Department of Development and Planning
Aalborg University
Fibigerstraede 11
DK-9220 Aalborg, Denmark
Telephone: +45 96 35 8338
Fax: +45 98 15 3537
E-mail: lbo@i4.auc.dk
http://Web.i4.auc.dk/lbo

Patrice Boursier
University of La Rochelle
L3i, Av. Marillac
17000 La Rochelle
France
E-mail: patrice.boursier@univ-lr.fr

António S. Câmara
Environmental Systems Analysis Group
New University of Lisbon
2825 Monte de Caparica
Portugal
Telephone: (351) (1) 2954464
E-mail: asc@mail.fct.unl.pt
http://gasa.dcea.fct.unl.pt

David DiBiase
Director, Deasy GeoGraphics Laboratory
Department of Geography
The Pennsylvania State University
University Park
PA 16802
USA
E-mail: dibiase@essc.psu.edu

Doris Dransch
Institute for Geodesy and Geoinformatics
University of Rostock
D 18051 Rostock
Germany
E-mail: dransch@agr.uni-rostock.de
http://Web.agr.uni-rostock.de/iggi/iggi.html

João Pedro Fernandes
Centro Nacional de Informação Geográfica
Rua Braamcamp, 82, 5° Esq.
1200 Lisbon
Portugal
Telephone: (351) (1) 386 00 11
E-mail: jpf@cnig.pt

Francisco C. Ferreira
Environmental Systems Analysis Group
New University of Lisbon
Quinta da Torre
P-2825 Monte de Caparica
Portugal
Telephone: +351-1-295-4464
E-mail: ff@mail.fct.unl.pt
http://gasa.dcea.fct.unl.pt

Alexandra Fonseca
Centro Nacional de Informação Geográfica
Rua Braamcamp, 82, 5° Esq.
1200 Lisbon
Portugal
Telephone: (351) (1) 386 00 11
E-mail: xana@cnig.pt

Pedro Gonçalves
Environmental Systems Analysis Group
New University of Lisbon
2825 Monte de Caparica
Portugal
Telephone: +351-1-2954464
E-mail: pmg@uninova.pt
http://gasa.dcea.fct.unl.pt

Cristina Gouveia
Centro Nacional de Informação Geográfica
Rua Braamcamp, 82, 5° Esq.
1200 Lisbon
Portugal
Telephone: (351) (1) 386 00 11
E-mail: cgouveia@cnig.pt

Arnaud Guilloteau
SILOGIC
78 Chemin des Sept Deniers
31200 Toulouse
France
E-mail: arnaud@silogic.fr

Sylvie Iris
SILOGIC
78 Chemin des Sept Deniers
31200 Toulouse
France
E-mail: sylvie@silogic.fr

Menno-Jan Kraak
Delft University of Technology
PO Box 5030
2600 GA Delft
The Netherlands
E-mail: kraak@itc.nl

Donatas Kvedarauskas
University of Paris-Sud
LRI - URA 410 CNRS
91405 Orsay, France

Mathilde Molendijk
Department of Regional Economics
Free University of Amsterdam
De Boelelaan 1105
1081 HV Amsterdam
Netherlands
Telephone: 31-20-4446099
E-mail: mmolendijk@econ.vu.nl

Joaquim Muchaxo
Environmental Systems Analysis Group
New University of Lisbon
2825 Monte de Caparica
Portugal
Telephone: +351-1-2954464
E-mail: jm@uninova.pt
http://gasa.dcea.fct.unl.pt

Jorge Nelson Neves
Environmental Systems Analysis Group
New University of Lisbon
2825 Monte de Caparica
Portugal
Telephone: +351-1-2954464
E-mail: jnn@mail.fct.unl.pt
http://gasa.dcea.fct.unl.pt

Edmundo M. N. Nobre
Environmental Systems Analysis Group
New University of Lisbon
2825 Monte de Caparica
Portugal
E-mail: en@mail.fct.unl.pt

Ana Pinheiro
Environmental Systems Analysis Group
New University of Lisbon
2825 Monte de Caparica
Portugal
Telephone: (351) (1) 295 44 64
E-mail: acp@students.fct.unl.pt
http://gasa.dcea.fct.unl.pt

Jonathan Raper
Department of Information Science
City University
Northampton Square
London EC1V 0HB
UK
E-mail: raper@soi.city.ac.uk

Armanda Rodrigues
Centro Nacional de Informação Geográfica
Rua Braamcamp 82 5º Esq,
1200 Lisbon
Portugal
E-mail: armanda@cnig.pt

Teresa Romão
Environmental Systems Analysis Group
New University of Lisbon
2825 Monte de Caparica
Portugal
Telephone: 351-1-2954464 ext: 0104
E-mail: tir@uninova.pt
http://gasa.dcea.fct.unl.pt

Henk Scholten
Department of Regional Economics
Free University of Amsterdam
De Boelelaan 1105
1081 HV Amsterdam
Netherlands
Telephone: 31-20-4446099
E-mail: hscholten@econ.vu.nl

Michael J. Shiffer
Planning Support Systems Group
Department of Urban Studies & Planning
Massachusetts Institute of Technology
77 Massachusetts Avenue, Room 9-514
Cambridge, MA 02139
USA
Telephone: (617) 253–0782
Fax: (617) 253–3625
E-mail: mshiffer@mit.edu

Predrag Sidjanin
Delft University of Technology
PO Box 5030
2600 GA Delft
The Netherlands

João P. Silva
Environmental Systems Analysis Group
New University of Lisbon
2825 Monte de Caparica
Portugal
Telephone: +351-1-2954464
E-mail: jps@uninova.pt
http://gasa.dcea.fct.unl.pt

Gerda Smets
Delft University of Technology
PO Box 5030
2600 GA Delft
The Netherlands

Maria Ines Sousa
Environmental Systems Analysis Group
New University of Lisbon
2825 Monte de Caparica
Portugal
Telephone: (351) (1) 295 44 64
E-mail: iss@students.fct.unl.pt
http://gasa.dcea.fct.unl.pt

Carmen Voigt
Institute for Geodesy and Geoinformatics
University of Rostock
D 18051 Rostock
Germany
E-mail: voigt@agr.uni-rostock.de
http://Web.agr.uni-rostock.de/iggi/iggi.html

Spatial Multimedia

CHAPTER ONE

Multimedia GIS: concepts, cognitive aspects and applications in an urban environment

Ralf Bill, Doris Dransch and Carmen Voigt
Institute for Geodesy and Geoinformatics, University of Rostock, D 18051 Rostock, Germany

1.1 CONCEPTS

The term 'multimedia', it seems, is on everyone's lips. In Germany, multimedia was the word of the year in 1995. Multimedia is an irreversible trend in computing. Nevertheless, multimedia comes with a range of completely different meanings: convergence of different worlds, such as telephone, computer and television; products such as videoconferencing, interactive TV and on-line services; or the integration of different media, such as text, graphics and video for information presentation. Multimedia enhances the overall quality and quantity of information. Most areas of information technology are embracing the emerging multimedia platforms, data types and environments. Multimedia may be seen as a concept that includes the technical as well as the application-oriented dimension of media integration. In the context of GIS, multimedia systems can be applied as information and monitoring systems, as learning systems or for co-operative work.

A multimedia system may be characterised as a computer-based system for integrated processing, storage, presentation, communication, creation and manipulation of independent information from multiple time-dependent and time-independent media (translated from Steinmetz et al., *1990).*

The major keywords in this definition are independent, integrated, time-dependent and time-independent. Each of these media should be treated independent of the others. They are integrated with the aid of computers and software, which should be able to synchronize the various sources according to spatial, temporal and content-specific relations. Integration of multimedia in a user's environment means that the user is not only viewing multimedia information but also creating and authoring multimedia objects. Time-independent media are text, graphics, tables, images and others. Time-dependent media are video and audio sequences, and sound. These media serve as information types in a multimedia system.

The idea of multimedia systems is a fairly old one, but the progress of hardware and software technology has brought this idea alive. Multimedia is extending in two major ways. On the one hand, multimedia systems are making use of the architecture and semantics of today's commercially available hypertext products. On the other hand, additional hardware such as video discs and real-time data compression boards are allowing the inclusion of time-dependent media, which were not part of hypertext systems.

Combining the above given definition of multimedia with a definition on GIS (Bill, 1994; Bill and Fritsch, 1994) leads to the following term:

A multimedia GIS may be characterized as a computer-based system consisting of hardware, software, data and applications allowing for integrated digital capture and editing, storing and organizing, modelling and analysing, presenting and visualizing spatially referenced data of multiple time-dependent and time-independent media.

The first GIS products with multimedia extensions are on the market. Craglia and Raper (1995) name two different approaches, which nowadays have to be extended to give a third approach:

1. 'GIS in Multimedia', integrating further data types and spatial analysis functionality in proprietary multimedia authoring tools, such as Toolbook, Macromedia Director and Authorware.
2. 'Multimedia in GIS', increasing the multimedia handling capabilities of existing GIS packages, such as MapInfo, ArcView and WinCAT.
3. 'World Wide Web plus GIS extensions', making use of the WWW functionality and its ability to integrate other types of software (e.g. Map Objects Internet Map Server, ArcView Internet Map Server, CGI scripts and Java applets)

Currently, no standard method exists for multimedia GIS implementations. With further standardization activities, such as Open GIS consortium, Open Doc and Windows functionalities (Object Linking and Embedding OLE 2.0, Component Object Model (COM) and Common Object Request Broker Architecture CORBA), new methods for implementation are appearing.

For a multimedia GIS, it is important to decide which media may support a specific task and how to make use of this in a daily computer environment. Current GIS technology mainly deals with time-independent media such as text, vector and raster graphics, and tables. The questions are as follows:

- Which time-dependent media are to be foreseen in a GIS?
- How can these media be combined with the existing time-independent media?

To answer this, investigations in the applications field as well as in human cognition are required (see chapter 2).

A major feature of a GIS is its analytical capability—an area in which many questions arise with respect to these new media:

- What type of analysis functionality is needed for new media such as images, video and audio, and what is the practical usage in a GIS environment?

- Which operations are suitable and meaningful for the new media?

An easy way to realize multimedia components is to integrate video or animation sequences of a planned situation in a GIS. A video sequence of the surrounds of a planned situation could help to decide a better planning alternative. A generated animation of a certain scenario, for instance, in environmental applications, may be helpful in answering 'What happens if?' questions.

Some specialized GIS deal with certain aspects of multimedia systems. For instance, in car navigation systems, time-dependent media such as a Global Positioning System (GPS) and an Inertial Navigation System (INS) are merged with selected spatial data on traffic routes and road signs in vector or raster format. Thus, position computation and route-finding can be combined with additional static or dynamic traffic information, such as traffic light sequences and vehicle volumes at a specific road segment. A further example of integrating time-dependent media in GIS is the TV inspection of sewerage systems, where the video-camera movement is georeferenced to the linear duct segment.

An interesting application for video may be an image sequence analysis—one that follows a spatially related object and derives a trajectory of its movement. Pure visualization techniques may also be useful, especially if they are context-sensitive. With the help of new media, the abstract view in GIS may become more realistic. The overlay of existing or planned information with video sequences may be one of the techniques used here. GIS will participate in the field of virtual reality.

1.2 COGNITIVE ASPECTS

In a multimedia GIS, it is necessary to decide which medium may support a specific task. To answer this, application fields as well as human cognition and function of media in the visualization process have to be considered. The application of media in different working fields has already been discussed (Cassettari *et al.*, 1993; Fonseca *et al.*, 1993, 1995a,b; Polydorides *et al.*, 1993; Blat *et al.*, 1995; Kerekgyarto *et al.*, 1995; Raper *et al.*, 1995; Câmara, 1997; Shiffer, 1997). In this contribution we want to point out cognitive aspects which are significant in a multimedia GIS.

Media in GIS have to show information about spatial objects and their relations, about spatial processes, and about planned situations and calculated scenarios. Multimedia techniques offer the whole palette of different media to present this information for investigation as well as for demonstration. The greater the palette of possible media, the more important it is to make the right choice. The following aspects can serve as guidelines for applying different media in a GIS.

The first aspect is the purpose that the medium has to fulfil. Should it give a vivid picture of an object or should it help to create a complex mental model from objects and their relationships? Cognition science (Weidenmann, 1991; Dreher, Mack 1996) distinguishes four types of functions that media may have:

- the function of demonstration

- the function of putting into context
- the function of construction
- the function of motivation

Each of these functions can be performed best by a specific type of medium.

1.2.1 The function of demonstration

Media for demonstration should help the user to get a suitable 'picture', a correct and complete idea of a phenomenon. Pictures, videos, realistic graphic representations and animations, as well as virtual reality, are best suited for this task. They can show the individual characteristics of an object. In the field of the urban environment, for instance, animations of planned buildings or of historical situations can improve the imagination and influence decision-making. Pictures of important or extraordinary buildings enclosed within a city map support peoples' navigation through the town. Also, audio can be applied in the above-mentioned way to present the sound of an individual place. For example, audio gives a vivid expression about the sound and loudness of a planned road with or without noise protection (see Figure 1.1). In particular, people who do not have a great knowledge of the topic need demonstration media.

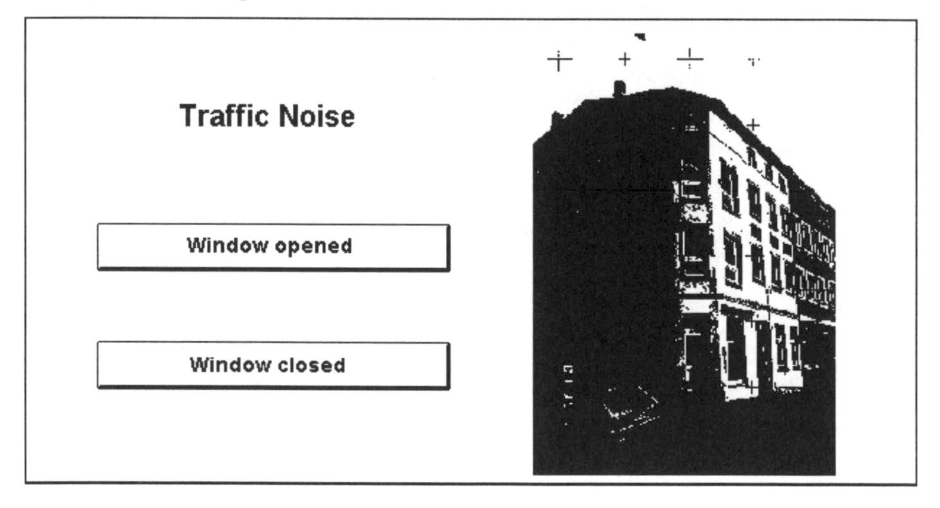

Figure 1.1 Traffic noise.

1.2.2 The function of construction

Media with the function of construction should help the user to create complex mental models. Mental models are constructions of knowledge about single information units and relationships. Media for this purpose have to inform about elements, their relations and co-operation. Pictures or realistic presentations are not suitable in this context. On the contrary, this function requires abstract media with prepared information such as maps, diagrams, graphs or formal sound. The creation of these models is highly influenced by the applied media. Graphs and diagrams should be used to present information on quantitative and qualitative relations, such as the structure of a population. Maps are to be applied to show spatial relationships. Dynamic presentations, such as animations, are capable of showing spatial processes such as the growth of a city. For example, creating a mental model of daily traffic flow in a city works best with a map or an animation of maps that show the traffic's spatial and temporal distribution. Videos and pictures can illustrate the phenomenon, but they can hardly create the mental model. Media for construction are to be applied in a GIS especially when analysing and exploring spatial data.

1.2.3 The function of putting into context

Media with this function should help the user to put information into a greater context. Media showing a wide spatial area, such as satellite images or video, can create a spatial context. Written or spoken text can give additional information, and media such as maps and diagrams can show comparable data from other areas or time periods. Sound is also able to put information into context. For instance, playing a typical sound of a particular area or time period (such as music or natural sound) supports the user's ability to identify and position the given information.

1.2.4 The function of motivation

Media with a motivational function should arouse the user's interest and attention. For this purpose, attractive pictures as well as dynamic media such as animation and video are best suited. Well known examples are 'fly throughs' that allow the user to 'fly' over a particular area.

In addition to these principles, some other aspects for media application in a GIS result from human cognition processes:

- to consider short-term memory's limited cognitive capacity
- to increase important information
- to avoid the overload of a single sense
- to support double encoding of information

1.2.5 Consider short-term memory's limited cognitive capacity

Human short-term memory has the limited ability to keep only seven information units simultaneously. If these information units are extended by repetition or elaboration, they are directed to the long-term memory; otherwise, they are forgotten. As a consequence of this limitation, media in a multimedia GIS should not offer to much information at the same time. This is particularly important when presenting different types of media simultaneously or when using dynamic presentations (such as animation or video) which transmit a lot of information within a very short time. For this reason, dynamic and multiple presentations should only be used if they actually support human cognition; for instance, visualizing a dynamic process by a dynamic animation. Otherwise, static and elementary presentations are to be preferred.

1.2.6 Increase important information

As mentioned above, multiple representations can overcharge the human cognitive capacity if they are not applied in a correct way. On the other hand, they also have the potential to improve information processing by emphasizing information. The different media in multimedia GIS, such as maps, pictures, text and sound, show various aspects of a spatial object or phenomenon. They increase the information about an object and accentuate it. Therefore information with great importance should be presented in a GIS not only by one medium but by a suitable combination of different media.

1.2.7 Avoid the overload of a single sense

A further aspect in a multimedia GIS is to avoid an overload of a single sense, especially the visual sense. Complex information has to be divided up between visual and auditive media to relieve single senses. For example, in car navigation systems a voice, instead of or in addition to a map, gives information about the driving direction. A further application is the exploration of multidimensional data sets, where a combination of visual and auditive media gives insight into complex data, but does not overload the visual sense.

1.2.8 Support double encoding of information

Double encoding of information means storing information in pictorial as well as in textual form in human memory. It is part of human cognition and should be supported by suitable media. In particular, persons who do not have a great knowledge of a subject will require pictures in combination with text to accomplish the double encoding process. On the other hand, experts will already have developed double encoded information; they will prefer single presentations, such as text *or* pictures. Media application in a multimedia GIS has to consider this aspect. For instance, GIS for education or tourism have to support double encoding; therefore, they have to present information in pictorial and textual form, such as maps or pictures, in combination with written or spoken text.

The outlined principles for media application in a GIS have been investigated in a first pre-test during a research project at Rostock University. In this exercise, a group of test persons had to deal with nine examples of various media and media combinations. The examples were derived from different fields of application and had to fulfil different communication functions. They included, for instance, presentation of noise along a planned highway for a citizens' meeting, presentation of a foreign city for orientation by foot and by car, and presentation of air quality during a certain period of time for a specialist's analysis, or for presentation to interested persons. In addition, the different media were applied in a learning system for GIS to support learners' cognition and information processing. A brief introduction to GIS, as well as a lesson for the travelling salesman problem, were realized. An evaluation of the learning programme, especially of the media used, as well as the pre-test confirm the introduced principles for media application.

1.3 APPLICATION IN AN URBAN ENVIRONMENT

Multimedia GIS have to manage, process and present time-dependent as well as time-independent media in one system. MM-GIS that allow a real connection between multimedia and GIS do not exist at the present time. For instance, search within the video and further reference of geoinformation is not yet realized. At the moment, links between the video and the map go no further than a click on the map, with the result that the video is played. The link between time-dependent media and GIS is currently the subject of research in a project at Rostock University.

The current investigation concerns how to realize the link between individual frames of a video and thematic data in an optimal way, so that a search of the database becomes efficient. By saving the link between frame and data, the database can be searched for certain terms and the related sequence can be played. The next step is to find objects within the video ('look for all sequences in which a tree is to be seen') and to link this information with a digitized map. Also, the spatial relations between objects have to be realized. Questions such as 'Which house is to the left of me?' or 'Which house is across the road from me?' could be answered within a video sequence. Further possibilities could allow the user to stop a video at any position, at which thematic information and geographical references would be given.

The prototype for a GIS-video link has been realized for a particular area in Rostock, where an extensive amount of data already exists. The data are, for instance, digitized maps, which include the geometry of objects (streets, park areas, single trees, areas with buildings, a playground or a tram), a digital elevation model, thematic information, a number of pictures, thermal and satellite photos, and a detailed 3-D model. A linked database contains further information (number of floors, type of use), and spatial data. In addition, a 20 minute video is available for this work on time-dependent media. It features a street located in the test area. The video was recorded with a camcorder and then digitized in order to allow further processing with existing software. The video was split into several sequences of different lengths (one for each building). The split video makes it easier to answer topological questions. Audio is included in the video and is available for further work. The software applied for the prototype is ArcView, because it offers the best basis for realizing the project; besides GIS functionality, ArcView provides its own script language, Avenue.

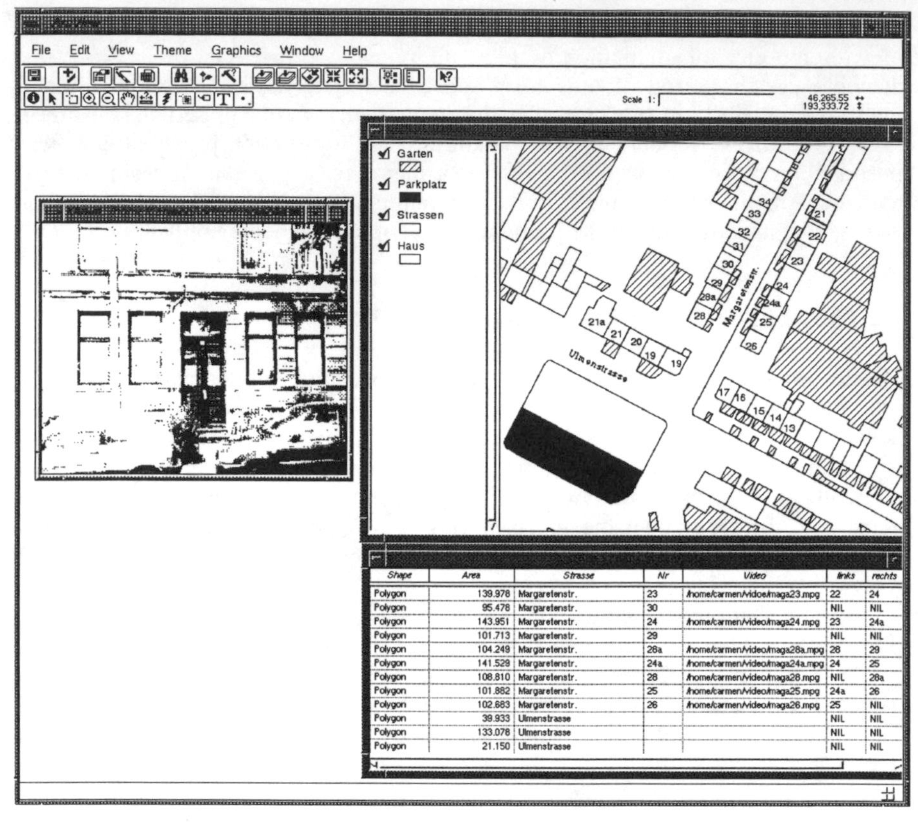

Figure 1.2 ArcView linked to video.

In the prototype, the topological relations between individual buildings and the related video sequences were inserted into the logical data model of ArcView. For instance, the object 'building' includes topological references to the neighbouring buildings on its left or right, or to the one across the street. In the case of a selective object analysis, the video switches to the neighbouring building and shows a picture of it. These functions are programmed using Avenue and the public-domain program 'xamin', which is integrated by system call. This is necessary because ArcView does not provide functions for processing time-dependent media. ArcView only enables the user to watch the video by stopping it at any point, as well as forwarding and rewinding the video. The first efforts to link video and GPS data in order to create local relations between objects automatically are under way. Here, the connection between individual frames of a video and the data of the real-time GPS measurement are examined, including a time component. With this technique, a video sequence for a defined area (e.g. video sequences within a radius of 50 m around a particular place) can be selected and presented.

1.4 CONCLUSION AND PERSPECTIVES

The idea of multimedia systems for GIS applications is fairly new. Major financial obstacles remain with regard to hardware and software. Although the systems are theoretically well developed, practice is still in its infancy, while standardization of tools and of the environment is totally lacking. A further obstacle is the specialization of GIS packages. It is fairly difficult to extend these systems to cope with additional media. Yet the potential of multimedia applications is enormous and future developments will clearly move them in this direction. In future, the GIS user may look somewhat different. Instead of sitting in front of a standard computer, he may walk in cyberspace. His virtual reality will be a combination of an abstract image of reality—the GIS database—plus planned scenarios and versions of what might become part of reality in future. The user will immediately see the impact of planning and he may interact with it and make alterations. New equipment will replace the standard tools that we use today. The user may wear a body suit and electro-optical sensors; a data glove will replace the mouse. Multimedia will bring a new dimension to GIS.

REFERENCES

BARTEL, S. and KÖNINGER, A., 1997, 3D-GIS for urban planning: object hierarchies, methods and interactivity, *Proceedings of the Joint European Conference on GIS*, Vienna, pp. 718–27.

BILL, R., 1994, Multimedia-GIS–definition, requirements and applications, in *The 1994 European GIS Yearbook*, GeoInformation International, pp. 151–4.

BILL, R. and FRITSCH, D., 1994, *Grundlagen der Geo-Informationssysteme*—Hardware, Software und Daten, Wichmann: Karlsruhe. 2. Auflage. 423 Seiten.

BILL, R., 1996, *Grundlagen der Geo-Informationssysteme*—Analysen, Anwendungen und neue Entwicklungen, Wichmann: Karlsruhe. 463 Seiten.

BLAT, J., DELGADO, A., RUIZ, M. and SEGUI, J. M., 1995, Designing multimedia GIS for territorial planning: the ParcBIT case, *Environment and Planning B: Planning and Design*, **22**, 665–78.

CÂMARA, A. S., 1997, Virtual reality presentation at the workshop spatial multimedia during Joint European Conference and Exhibition on Geographical Information Systems, Vienna.

CASSETTARI, S. and PARSON, E., 1993, Sound as a data type in a spatial information system, in *Proceedings of European Conference and Exhibition on Geographical Information Systems*, Vol. 1, S. 194–201, Genoa.

CRAGLIA, M. and RAPER, J., 1995, Guest editorial: GIS and multi-media, *Environment and Planning B: Planning and Design*, **22**, 634–6.

DRANSCH, D., 1997, Cognitive aspects applied to computer based GIS learning systems, *Proceedings of the Joint European Conference on GIS*, Vienna, pp. 1291–9.

DREHER, A. and MACK, J., 1996, Medienpsychologische Grundlagen, Skript zur Vorlesung Mediendidaktik M12, Prof. Dr. Kerres, http://mi-hp00.mi-lab.fh-furtwangen.de/md/md4-1.html.

FONSECA, A., GOUVEIA, C., RAPER, J., FERREIRA, F. C. and CÂMARA, A. S., 1993,

Adding video and sound to GIS, in *Proceedings of European Conference and Exhibition on Geographical Information Systems*, Vol. 1, S. 187–93, Genoa.

FONSECA, A., GOUVEIA, C., CÂMARA, A. S. and SILVA, J. P., 1995a, Environmental impact assessment with multimedia spatial information systems, *Environment and Planning B: Planning and Design*, **22**, S. 637–48.

FONSECA, A., GOUVEIA, C., FERNANDES, J. C., CÂMARA, A., PINHEIRO, A., ARAGAO, D., SILVA, J. P. and SOUSA, M. I., 1995b, Expo'98 CD-ROM, A multimedia system for environmental exploration, in *Proceedings of First Conference on Spatial Multimedia and Virtual Reality*, S. 139–49, Lisbon.

FONSECA, A., GOUVEIA, C., CÂMARA, A. and SILVA, J. P., 1995, Environmental impact assessment with multimedia spatial information systems, *Environment and Planning B: Planning and Design*, **22**, 637–48.

GROOM, J. and KEMP, Z., 1995, Generic multi-media facilities in *Geographical Information Systems*, in FISHER, P. (Ed.), *Innovations in GIS*, Vol. 2, London: Taylor & Francis, pp. 189-200.

GROSS, R. and WILMERSDORF, E., 1997, New GIS gateways for the citizens: multimedia and Internet services, in *Third European Conference and Exhibition on Geographical Information*, IOS Press, pp. 1499-508.

GRUBER, M. and WILMERSDORF, E., 1997, Towards a hypermedia 3D urban database, in *Third European Conference and Exhibition on Geographical Information*, IOS Press, pp. 1120-8.

HEYWOOD, I., BLASCHKE, T. and CARLISLE, B., 1997, Integrating geographic information technology and multimedia: the educational opportunities, in *Third European Conference and Exhibition on Geographical Information*, IOS Press, pp. 1281-90.

KEREKGYARTO, E., 1995, Multimedia application as input of GIS database, in *Proceedings of Joint European Conference and Exhibition on Geographical Information Systems*, The Hague.

KHOSHAFIAN, S. and BAKER, A. B., 1996, *Multimedia and Imaging Databases*, San Francisco: Morgan Kaufmann.

MORENO-SANCHEZ, R., MALCZEWSKI, J. and BOJORQUEZ-TAPIA, L. A., 1997, Design and development strategy for multi-media GIS to support environmental negotiation, administration, and monitoring at the regional level, *Transactions in GIS*, **1**, 161-75.

POLYDORIDES, N. D., 1993, An experiment in multimedia GIS: great cities of Europe, in *Proceedings of European Conference and Exhibition on Geographical Information Systems*, Vol. 1, S. 203-12, Genoa.

RAPER, J. and LIVINGSTONE, D., 1995, The development of a spatial data explorer for an environmental hyperdocument, *Environment and Planning B: Planning and Design*, **22**, 679-87.

SHIFFER, M., 1995, Interactive multi-media planning support: moving from stand-alone systems to the World Wide Web, *Environment and Planning B: Planning and Design*, **22**, 649-64.

SHIFFER, M. J., 1997, A geographically-based multimedia approach to city planning, Paper presented on the European research conference on socio-economic research and geographic information systems, Pisa.

STEINMETZ, R., RÜCKERT, J. and RECKE, W., 1990, Multi-Media-Systeme, *Informatik Spektrum*, **13**(5), S. 280-2.

WEIDENMANN, B., 1988, *Psychische Prozesse beim Verstehn von Bildern*, Bern.

CHAPTER TWO

Magic Tour: integrating multimedia and spatial data in an authoring system for tourism

Patrice Boursier[1] and Donatas Kvedarauskas
University of Paris-Sud, LRI - URA 410 CNRS, 91405 Orsay, France
and
Sylvie Iris and Arnaud Guilloteau
SILOGIC, 78 Chemin des Sept Deniers, 31200 Toulouse, France

2.1 INTRODUCTION

Tourism applications are of major interest for tour operators, tourist offices and travel agencies. In the framework of the ESPRIT Project 8752—*Magic Tour*, major emphasis has been devoted to the development of a multimedia authoring system specifically oriented to developing tourism-oriented applications. This is the reason why *Magic Tour* is not only a multimedia authoring system, but it also provides GIS features that make it different from other similar tools available on the market.

In particular, *Magic Tour* is an authoring system for supporting the generation of a wide spectrum of tourism applications based on multimedia and Geographical Information Systems (GIS) technologies.

Through the adoption of GIS technologies, new classes of operations based on adjacency, distance, proximity calculus and route optimization are made available to the users of tourism applications, in addition to more traditional multimedia data navigation and presentation features.

The *Magic Tour* project started in September 1994 and ended in February 1997. It has produced two different kinds of results:

1. An authoring environment for building tourism applications that integrate multimedia and geographical data.
2. Three prototype applications developed to validate the authoring system and to

[1] Present address: University of La Rochelle, L3i , Av. Marillac, 17000 La Rochelle, France.

support different classes of users.

The functional requirements of the authoring system are described in section 2.2, along with basic map manipulations. Architectural considerations are presented in section 2.3. The overall results of the *Magic Tour* project are briefly discussed in section 2.4.

2.2 FUNCTIONAL REQUIREMENTS AND MAP DATA HANDLING

The tourism applications aimed at fall into three categories that depend upon different classes of potential users:

- tour operators, for the preparation, updating and consultation of packages and catalogues
- tourist offices of a country, a region or a city, so as to offer multimedia and geographical information regarding various tourist resources
- travel agencies that will, in addition, propose various services and solutions to potential customers

Since we intended to develop an authoring system, two different classes of users have been considered.

The *primary users*, or *authors*, will use the authoring system to produce multimedia applications in order to promote their products. They are mainly tour operators, for construction, updating and consultation of tourist packages. They are also tourist offices, so as to offer the visitor integrated and geographically relevant information regarding tourist resources, transport and routing to support the visitor's interactive free choice.

The *final users*, who are tourists, will use the final applications in order, for example, to seek out a destination according to their budget and specific interest, set a travel itinerary, and organize their visit around a city or a region. Travel agents are also users, so that they can show the customer the different solutions and make him capable of realizing his own specific targets.

Consequently, two levels of requirements have been identified, the authoring system level and the application level respectively.

At the *authoring system level*, the authoring system is dedicated to users that are not familiar with computers. Ease of use is then a major requirement, together with low cost. The other identified requirements are:

- flexibility, modularity and extensibility
- standards, multilingual
- network transparency
- interaction with other applications, database management systems (DBMSs), geographical information systems (GISs) and reservation systems
- the ability to generate off-line applications (e.g. on CD–ROM)

At the *application level*, the general requirements are as follows:

- navigation capabilities
- activation of multimedia elements
- identification and selection of subjects of interest, e.g. museums or hotels
- route editing and selection with possible constraints, e.g. time, distance or places to visit
- reservation through gateways to products such as Galileo or PC Travel

Among the identified requirements, some need specific GIS functionalities, mainly:

- preparation of maps: this function, which includes data import and editing, must be provided at the authoring level
- display of maps
- processing and answering information queries based on geographical identification and subsequent searching for related information
- dynamic routing to calculate the best itinerary, taking into account the user's preferences

The *Magic Tour* toolbar appears in Figure 2.1. It consists of a set of buttons and menus. From left to right, the buttons allow the author of a *Magic Tour* application to edit texts, select objects (frames, buttons, etc.) for editing and manipulation, create frames containing texts and images, create drawing areas, create movie frames, create map frames and manipulate frames in general. The last two buttons correspond to selection of the *author mode* (hammer) and the *browser mode* (book) respectively.

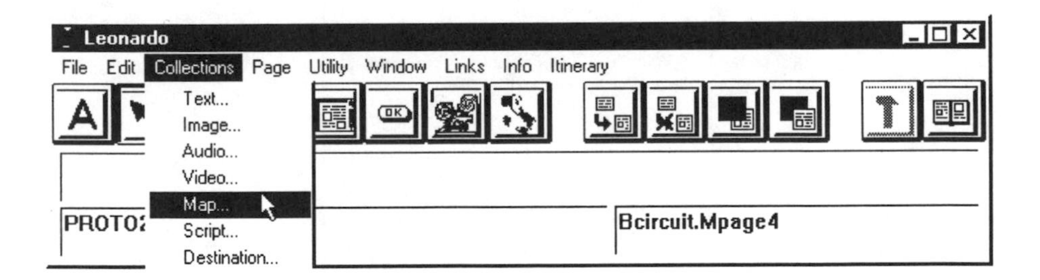

Figure 2.1 The Magic Tour toolbox and menus.

Before building applications, *Magic Tour* authors must import objects of different types into *collections*. This is illustrated by Figure 2.2, where the map collection is selected.

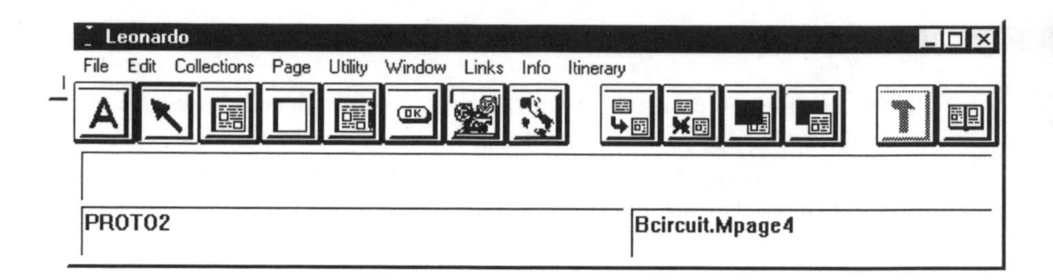

Figure 2.2 Map collection selection.

Figure 2.3 shows the different possibilities offered for handling maps and the way in which they can be previewed by the author.

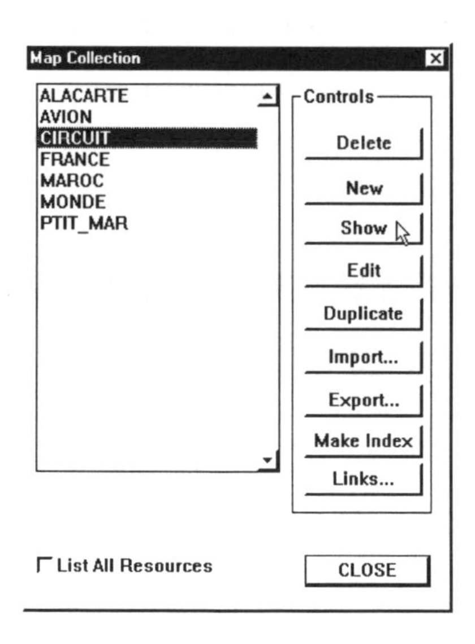

Figure 2.3a Map collection handling and map preview.

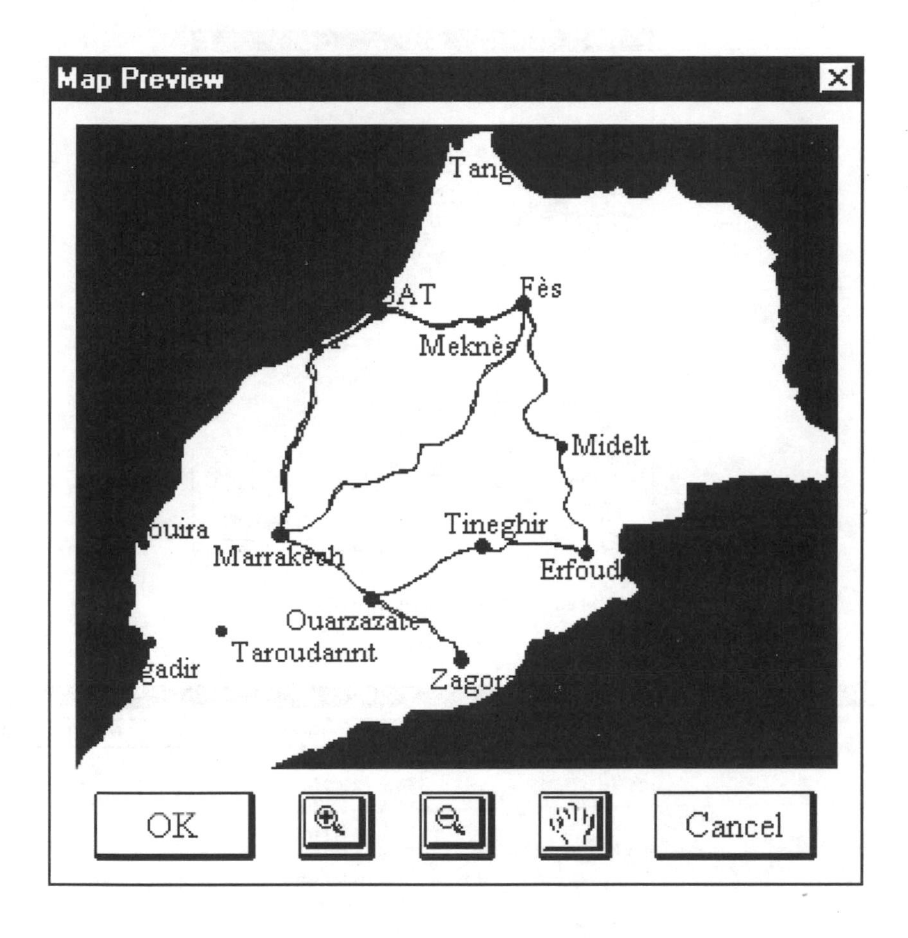

Figure 2.3b Map collection handling and map preview.

Map editing is shown in Figure 2.4, with an example illustrating itineraries. The final result of this work appears in Figure 2.5.

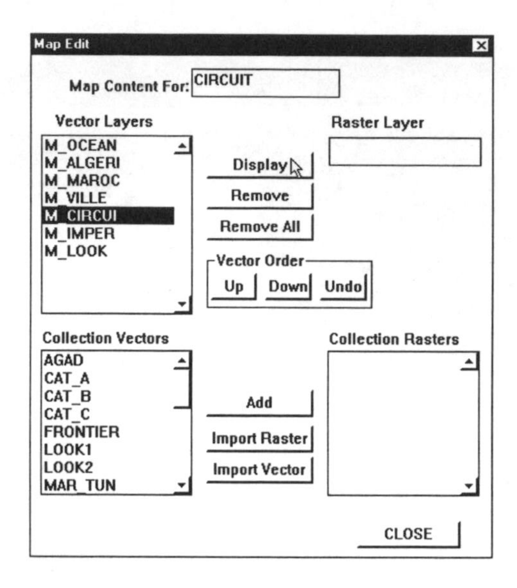

Figure 2.4a Map editing with itineraries.

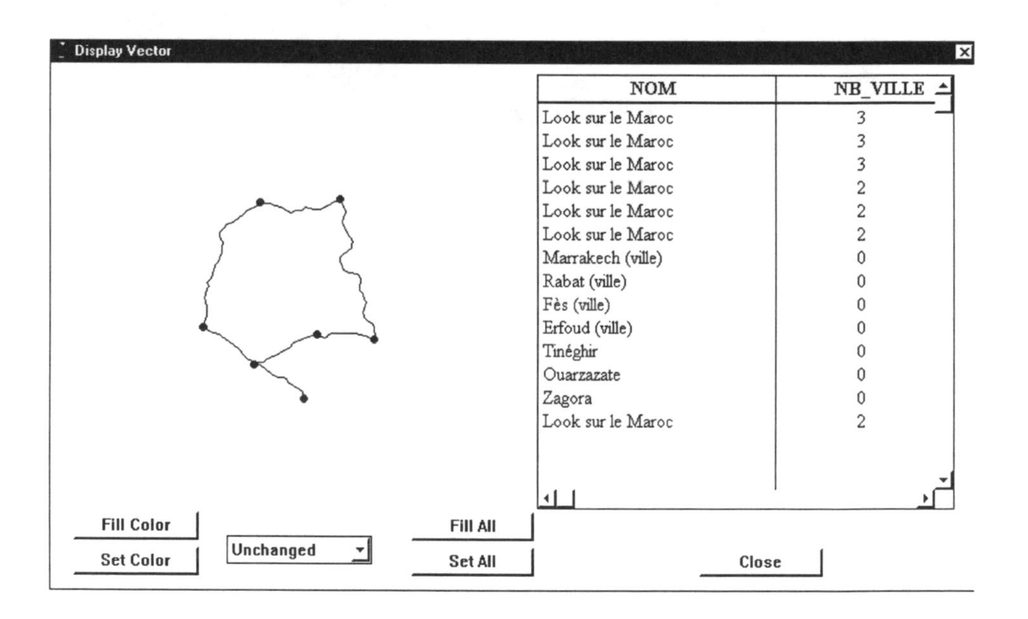

Figure 2.4b Map editing with itineraries.

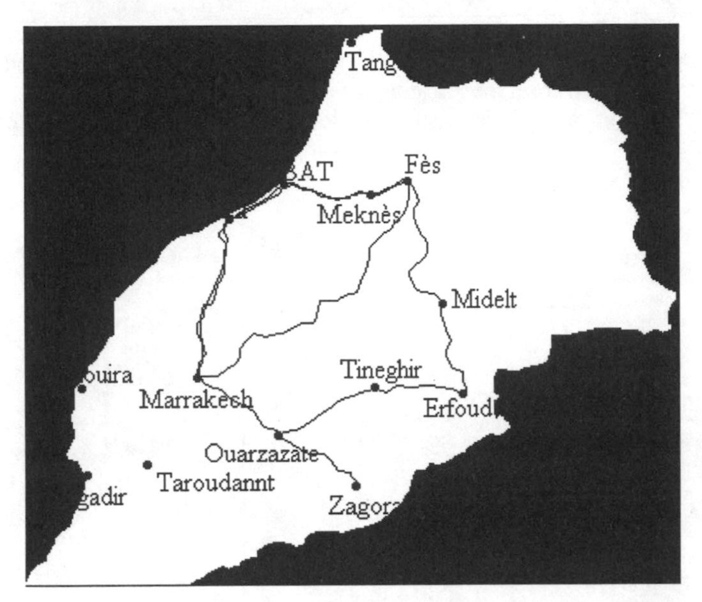

Figure 2.5 A sample map with itineraries.

2.3 ARCHITECTURAL CONSIDERATIONS

Several basic approaches have been considered in order to provide the required functionalities in the authoring system.

They mainly consist of:

a extending an authoring system with required multimedia and GIS functionalities;

b extending a GIS tool with required multimedia functionalities and hypermedia capabilities.

Another possible solution that we have adopted consists of integrating existing products and developing extra functionalities in order to meet specific requirements.

The *Magic Tour* consortium has developed from scratch the basic GIS functionalities required to meet the user's needs. These basic functionalities include the preparation of more or less complex maps by superposing (or overlaying) raster and vector basic layers. For example, it is possible to superpose a point-based HOTELS layer and a line-based ROADS layer on to a raster-based image of a REGION, derived from the scanning of a paper map.

The solution that has been designed and implemented is summarized in Figure 2.6.

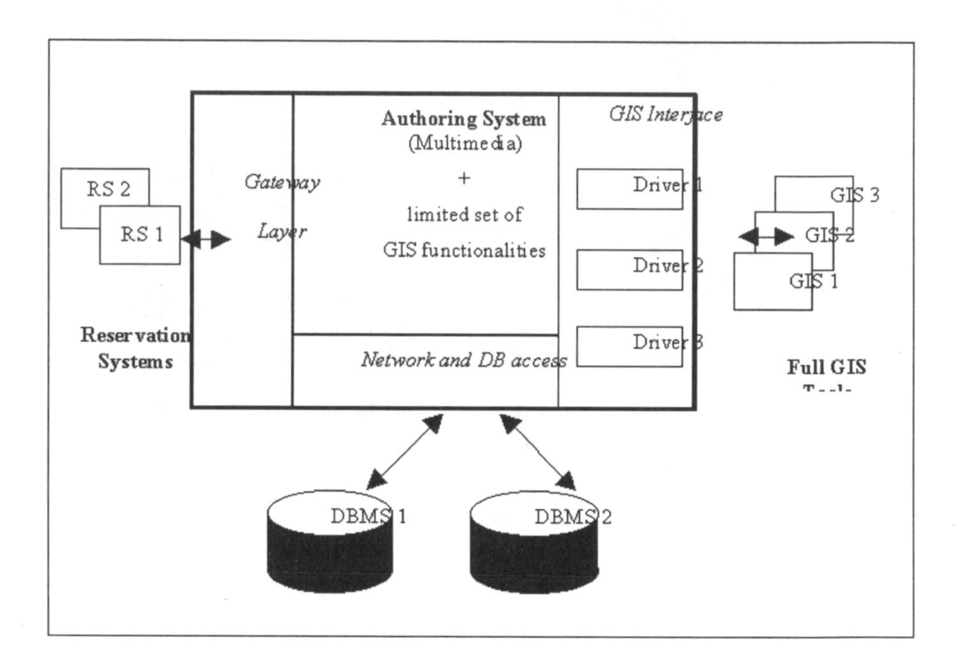

Figure 2.6 The Magic Tour authoring system general architecture.

We have also developed drivers to interface widely used market tools in order to provide more complete GIS functionalities. These functionalities mainly concern vector-based geographical data processing. The selection of a geographical feature indeed implies the handling of individual vector-based geographical entities, both from a data management and from a graphical point of view. This is a basic functionality of vector-based GIS tools, and we did not intend to redevelop it.

Consequently, only simple *spatial queries* that address individual entities are possible with the basic authoring system. More complex ones are only handled through an external GIS link.

The link to external GIS tools can be made in two different ways:

1 Using external libraries (DLL), which are made available by most existing products that run in the Microsoft Windows environment.

2 Opening the GIS environment directly from the authoring system environment using OLE technology. This second kind of link is not presently available in many PC-based GIS tools.

2.4 CONCLUSION

The *Magic Tour* project has produced the following main results:

1 The authoring system, which integrates multimedia and basic GIS functionalities.

2 Three prototype applications developed to validate the authoring system.

The three prototype applications have been based on the requirements of tour operators, tourist offices and travel agencies respectively (tourism-oriented partners in France, Italy and Greece).

The solution adopted for designing the authoring system consists of the extension of an existing authoring system with multimedia capabilities. Basic GIS functionalities have been included in the *Magic Tour* final product. More powerful GIS functionalities are made available using external links to commercial GIS tools.

Following *Magic Tour*, the *Magic Tour Net* project has been selected for funding by the European Commission within the framework of the INCO-DC programme. It will start in January 1998. *Magic Tour Net* can be seen as the evolution of *Magic Tour* to the Internet, in order to offer a product that allows the development of tourism-based applications in an Internet/Intranet environment.

Another major objective of *Magic Tour Net* is to give autonomy to the technology providers that reside in the developing countries, in offering new and advanced tools to the local tourism industry.

REFERENCES

BOURSIER, P. and MAINGUENAUD, M., 1992, Spatial query languages: extended SQL vs. visual query languages vs. hypermaps, in *Proceedings of the 5th International Conference on Spatial Data Handling (SDH'92)*, Charleston, VA, August.

BOURSIER, P., KVEDARAUSKAS, D., IRIS, S. and GUILLOTEAU, A., 1996, Integration of mutimedia and spatial data in an authoring system for building tourism applications, in *Proceedings of the International Conference on Information and Communication Technologies in Tourism (ENTER'96)*, Davos, Switzerland, Berlin: Springer-Verlag.

BOURSIER, P., KVEDARAUSKAS, D. and SPYRATOS, N., 1997, Integration of multimedia and GIS technologies, in *22nd International Conference on Information Technologies and Programming*, Sofia, Bulgaria, June.

DBOUK, M., *et al.*, 1996, Querying and visualizing geographical data sets with dynamic maps, in *Proceedings of the 2nd ACM Workshop on New Paradigms for Information Visualization and Manipulation*, Rockville, MD.

GARZOTTO, F., MAINETTI, L. and PAOLINI, P., 1995, Evaluation of hypermedia tourism applications, in *Proceedings of Information and Communication Technologies in Tourism International Conference (ENTER'95)*, Innsbrück, Austria, Berlin: Springer-Verlag.

KVEDARAUSKAS, D., *et al.*, 1997, GEOLIB as a software component for making GIS interoperable, in *1st International Conference on Interoperating Geographical Information Systems*, Santa Barbara, CA, December.

CHAPTER THREE

Future directions for hypermedia in urban planning

Lars Bodum

GISplan, Department of Development and Planning, Aalborg University, Fibigerstraede 11, DK-9220 Aalborg, Denmark

3.1 INTRODUCTION

The past 10 years have brought many new technologies to local planning offices (Bodum, 1997). The evolution of computers and specifically the speed of microprocessors have been very important factors during this period. But the hardware is only the means for the message. The way in which planners use computers and how they communicate through these new media are even more important than the speed of the hardware. Many people in the cartographic and GIS community have regarded geographical information technology as a set of tools, or as individual programs or functions for specific purposes. This will change in the near future. The different technologies will fuse together and the object-oriented approach to the development of specific applications will become even more evident (Shiffer, 1995; Batty, 1997b; Raper, 1997).

The best example of the integration of different technologies into several layers of media is the development of the World Wide Web (the Web). Here is one type of interface for all information. This type of hypermedia interface has become the standard for the distribution of geographical information, and both professionals and citizens in general will gradually have access to giant databases of maps, attributes and planning proposals (Bodum, 1995; Mitchell, 1995; Shiffer, 1995; Batty, 1997b; Raper 1997). It has been said that urban planners are among the professionals who will benefit from this evolution (Shiffer, 1995). This may be true if the urban planners are prepared for this transition, and if they have the kinds of messages that will fit into these new media. The message of today, in the form of planning proposals and reactions from citizens, does not immediately fit into the new media. It takes time to make this transition in working methods: data, documents and new possibilities for electronic communication have to be evaluated for use in the planning process. As we stand at the beginning of this new world, it is not yet possible to state in any categorical way how the media will be used at various stages of the planning process (Batty, 1997b). However, this chapter will try to give a few guidelines about what to look for in this new electronic world.

Recent surveys show that the Danish planners are not yet fully prepared for this change of media (Larsen *et al.*, 1998). Only 5% of the municipalities in Denmark have started to use the new media for communication of the master plan: 25% have serious intentions about using the media, but the rest of the municipalities (70%) have no plans to change the media that they use today. In 33% of the Danish municipalities, decisions have not yet been made about the use of the Internet for the communication of municipal plans (Larsen *et al.*, 1998). It is clear that lack of knowledge about the possibilities offered by the new media is one of the major reasons for the lack of interest. To change this situation, it becomes necessary to promote the use of new geographical information technologies, where the results are more suitable for electronic publishing; for example, in 3-D applications, distributed GIS, geodata libraries and networked communication systems. As long as paper still rules in planning offices, the arguments for change are too weak.

3.2 HARDWARE THEN AND NOW

What happened 20-30 years ago with respect to cartography and the handling of geographical information has been referred to as a revolution (Tomlinson, 1988). The introduction of computers in planning offices meant a total change, in many ways. This situation has been improved constantly ever since, because of the evolution of computer technology. With respect to the efforts made in the early days of automatic mapping, it should be mentioned that in those days computers were very expensive and the processing was very time-consuming. Tomlinson (1988) puts this into perspective in his article about the transition from analogue to digital cartographic representation. For the development of the CGIS (Canada Geographic Information System[1]), an IBM 360/65[2] (worth $12 million in 1988 dollars) with 16K of RAM and a performance around 0.5 MIPS was used. Compared to an ordinary PC of today, which runs with 32 MB of RAM and around 200 MIPS, it seems almost incredible that any serious automated mapping could be performed in those days. To scan maps into the system, a two-bit (black and white) 48 × 48 inch scanner was also developed, and was priced at $180.000: many other examples from this time will show the same pattern (Tomlinson, 1988). Even though the tools seemed very primitive and very expensive, the transition from analogue to digital techniques happened during these years (1960-1970) and this period has had a great deal of influence on the current level of the technology today. Despite all the constraints (e.g. very expensive, low-performing computers), this first wave of GIS and automated mapping tools showed that it was possible to produce new types of representations of the world based on new data models and new analytical methods. It was really a breakthrough in both the handling of geographical information and in visualisations within GIS, and it showed that it was possible to focus on the solutions and on the message. The

---

[1] The Canada Geographic Information System was the first system to use the abbreviation GIS, for 'geographic information system'.

[2] The IBM 360/65 was a very popular mainframe produced by the leading computer manufactor at that time. In 1964-65, IBM had an 80% share of the total computer market.

mainframes of the 1960s and the workstations of the 1970s were eventually replaced by PCs in the 1980s, and up into the 1990s this downscaling of computers is still under way. Lately, there has been a focus on laptops, Personal Digital Assistants (PDA) and Network Computers (NC), and if this trend continues we can look forward to wireless units that are comparable in size to cellular phones or portable CD players, and with the ability to connect to the network via orbiting satellites. The developments in hardware are more or less predictable. It is not possible to say the same about the solutions (software systems) that will run on these units.

3.3 THE NEXT WAVE IS DEDICATED HYPERMEDIA AND HYPERMAPS

From the mid-1980s to the present time, many planning offices have been implementing different kinds of information technology. The introduction of the PC has made this possible. First came word-processing, databases and spreadsheets, and after that followed automated mapping (CAD) and eventually GIS (Bodum, 1997). The situation today is that all of these very different applications have been integrated in solutions such as desktop publishing (e.g. PageMaker and Quark Xpress) and desktop GIS applications (e.g. ArcView, MapInfo and Atlas GIS), which are known for their ability to combine different types of data in one document. Following this concept, this means that GIS today is identified as one application running on a specific type of computer and producing one specific type of document. This means that you have to differentiate between operating systems, computer hardware and other connected peripherals when you talk about a specific GIS. GIS, as it is seen today, is one system on one computer. This is one of the consequences of the evolution of computers over the past 30 years.

The development of software for urban planning purposes is evolving under the influence of both multimedia and hypermedia concepts. Over the past 5–7 years, there have been a long line of interesting developments in this area, but very few of them have been successful in terms of implementation in real-life situations (Raper, 1997). Their purpose has been to explore the possibilities of digital media and to show alternative ways of communicating in urban planning. Raper (1997) reports the progress towards spatial multimedia and mentions several examples where hypermedia are used in connection with urban planning. Some of the most interesting projects during this period have been carried out by Shiffer (1992, 1995), with a specific focus on new techniques for building interfaces for different planning purposes, and for collaborative work in planning situations. The work of Shiffer and his colleagues at MIT shows that hypermedia consist of more than just clickable maps and that it is possible to design planning support systems which handle the human-computer interaction in a more intuitive way.

Another interesting project, called 'Townplanner', was developed by Kiib and Veirum (1993). 'Townplanner' (Figure 3.1) was a prototype of a hypermap application, and the idea was to create a virtual desktop where the planner could digitise, store and view all kinds of information related to a specific case. The objects (or perhaps 'items' would be better) related to the case were digitally represented by small icons on a 240 × 180 pixel control palette. A certain part of this small virtual desktop represented the visible screen displayed by the computer monitor. When items were dragged into this part of the control palette, they popped up on the monitor as documents. Every piece of information, every little item belonging to a case, was linked through the geographical reference (Bodum, 1997).

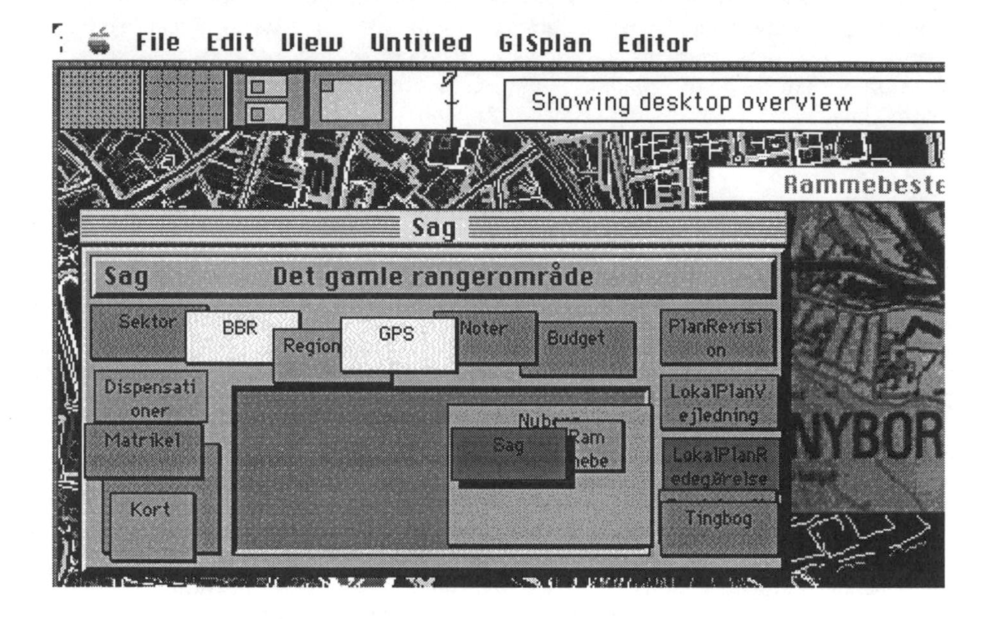

Figure 3.1 A screen dump of 'Townplanner'.

Because the area of hypermedia within geographical information science grew very rapidly at the start of the 1990s, there was a need for a more tight definition of this notion. The term 'hypermap' appeared in the literature. It was initially introduced by Laurini and Thompson (1992). A broader definition of the term is given in Raper (1997):

> When a hypermedia spatial database is integrated with coordinate-based spatial referencing such that each spatial 'object' has a stored location, the system can be defined as a hypermap.

The potential for hypermaps to become widely used depends on the ability of developers to conceive new spatial abstractions which can be implemented in the hypermap environment (Raper, 1997). This is the real challenge in the near future.

But although the term 'hypermap' was introduced in the 1990s by workers such as Laurini and Thompson (1992), there had been earlier attempts to utilise hypermedia in connection with the built environment. One of the very first examples of the use of this technology had already been shown in the 1970s, when scientists from the Massachusetts Institute of Technology (MIT) created the Aspen Movie Map. This project demonstrated an interactive application in which a map was combined with dynamic images stored on laser discs, and in which a user was able to move along the streets of Aspen, Colorado, and interactively navigate around the streets, by clicking on the desired direction of travel. It was even possible to click on a specific buildings and get information about it. Negroponte (1996) was one of the developers, and here he describes the system in his own words:

> In 1978 the Aspen Project was magic. You could look out your side window, stop in front of a building (like the police station), go inside, have a conversation with the police chief, dial in different seasons, see buildings as they were forty years before, get guided tours, helicopter over maps, turn the city into animation, join a bar scene, and leave a trail like ariadne's thread to help you get back where you started. Multimedia was born.

Some of these first initiatives have now been proved right because of the popularity of the Web. There is no doubt that the future of the hypermap has been secured by the development of the Web over these years, and the new spatial abstractions needed for future developments of the hypermap will be shown on the Web. The development of spatial multimedia/hypermedia solutions will become very important over the next 3–5 years, and the great focus on specific hardware and software for GIS in planning offices will eventually become less important. GIS will change from being a specific product in a box, to become a part of all the spatially referenced multimedia and/or hypermedia solutions.

But it does not stop there. The next generation of computer technology will soon be upon us, and this will bring forward the wireless-but-still-networked computer. This means that the individual processes can be shared between the computers (perhaps a better term would be 'units') that are available on the network when the service is needed. This will be possible because the code will be written in a common programming language such as Java, and the representation will be carried out in a common browser, like the ones that we see on the Web today. This will also bring back the object-oriented approach to deal with large geographical databases.

What can be learned from this history of evolution of computer technology is that the specific hardware and software for GIS has played a very important role in the first 30 years of automated mapping and GIS, but that this will change dramatically within the next 5 years. GIS has been, and still is, identified by the specific application and the specific hardware on which to run it. With the advent of networked computer technology, there will be a more dedicated role for the message and for the communication of the message (Fig 3.2).

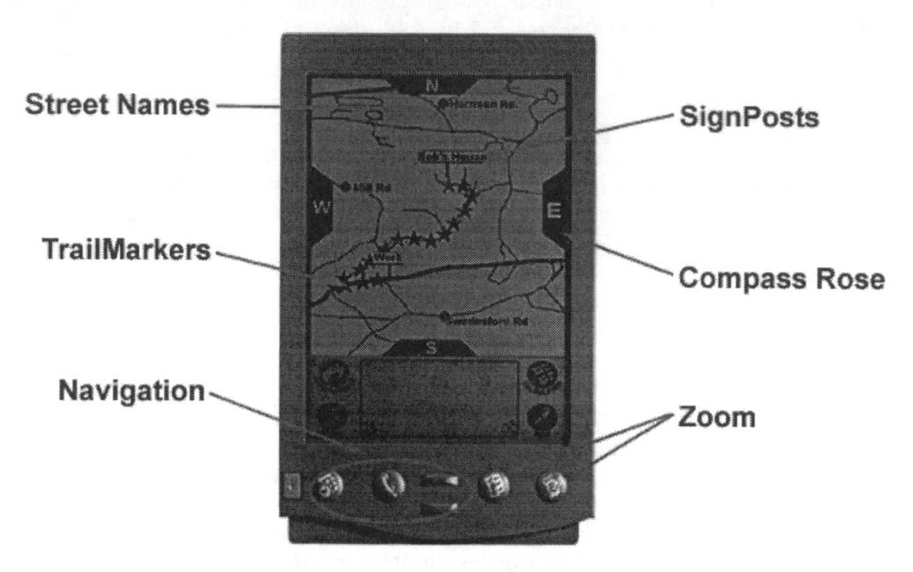

Figure 3.2 The PalmPilot (soon to be wireless-networked) with a map on the screen.

3.4 THE MAP AS CARRIER OF INFORMATION

To help in this process of transition it is very important to look at the map as the carrier of information, and it is also very important to find new ways to navigate through the huge piles of geographically orientated material that will be made available through the network. It is difficult to point out the dependent variable in this evolution, but it is a fact that the change within the areas of geography and cartography has been noticed, and that it has been referred to as another reformation of the technology:

> Currently we have reached a point where we could start referring to a third reformation in cartography. Where the second reformation revolutionised map production, keeping the cartographic medium (paper) the same, the third brings the virtual map that primarily serves the user on-screen and is supported by relatively complex user interaction allowing the user to take an active part in the map making process. (Koop, 1995)

Fisher (1997) also points out that, in some cases, the traditional metaphor for communication of geographical and cartographic information, the map, is outdated and must be changed in respect of the findings within thc ficld of visualisation:

> ... the use of the term 'map' to describe a part of a database (commonly a layer) remains endemic among GIS users. Perhaps it is a term in transition, but it seems to reflect the conceptualisation of the data as hard copy objects. The belief that the hard copy map is the ultimate product of a GIS (however produced) is also prevalent, in spite of the developments in scientific visualisation. (Fisher, 1997)

Marble (1990) also points out other problems with the traditional map, and how difficult it is to use it with the kind of complexity that exists in normal GIS applications. If GIS functions become available through the Internet and the only interface is the common browser for the World Wide Web, then it will become very difficult to clarify the meaning and the form of communication in general:

> Although maps appear at first sight as relatively simple iconic devices (so simple in fact that most societies do not even provide formal instruction in their use for many children), a major difficulty has been that the resultant documents have not lent themselves to easy use within the framework of more than the simplest attempts at data extraction. Complex queries, especially those which contain a quantitative component (such as multiple measurements of distance, direction or area), simply cannot be processed easily and quickly by the user even with the assistance of the simple tools available. (Marble, 1990)

It seems evident that the solution for the display of geographical information in the networked environments of the future, and with the use of multimedia and hypermedia applications, is not only the electronic map on a computer monitor. The map has to be combined with other kinds of navigational tools that are more usable in networked hypermedia environments. This is not the same as saying that the map is useless in this new world, anything but! The map will still be a very important element in future hypermedia applications for urban planning.

3.5 VIRTUALITY AND REALITY

In urban planning there is in many cases a need for the ability to mix the reality and the virtuality; or, in other words, to understand the visual effects from a specific spatial plan implemented in a local area you must try to imagine what the future physical elements, such as buildings, roads or trees, will look like and how they will be integrated in the area. This task can be carried out in many ways. First, for many years planners have used the ability to create beautiful perspective drawings and paintings, in which reality is mixed with planned objects. This colourful artwork is the outcome when planners use classical methods of illustration, which have been developed over several centuries. Secondly, they will perhaps build physical models that illustrate the terrain and existing buildings together with the buildings suggested in the plan.

There is at least one big disadvantage with these conventional methods. It is almost impossible to change the plan in an interactive way without creating a new illustration or a new physical model. The flexibility that you need when the conditions in the plan or the subject of the visualisation are changed is not available. Furthermore, the ability to show visualisations of more abstract geographical information in a 3-D perspective is not present either. This may not be a major problem for professional planners, because they have learned to interpret planning documents, but it is certainly a very tangible problem for politicians and for the ordinary citizens in the area of interest. They have to trust their own imagination, or the few illustrations and models that are produced on the basis of the information given in the planning documents. This can lead to potential risks with regard to democracy and public participation in the urban planning process.

3.6 URBAN PLANNING MOVES OUT OF THE FLATLANDS

The ability to work in more than two dimensions will eventually move urban planning away from the flatlands again. There is a need to visualise planning proposals in 3-D, and even in a dynamic way so that time, as the fourth dimension, also will become a possibility. Tufte (1990) expresses it this way:

> Even though we navigate through a perceptual world of three spatial dimensions and reason occasionally about higher dimensional arenas with mathematical ease, the world portrayed on our information displays is caught up in the two-dimensionality of the endless flatlands of paper and videoscreen. All communication between the readers of an image and the makers of an image must now take place on a two-dimensional surface. Escaping this flatland is the essential task of envisioning information, for all the interesting worlds (physical, biological, imaginary, human) that we seek to understand are inevitably and happily multivariate in nature. Not flatlands.

The combination of 3-D and hypermedia is very obvious. It will be possible to link specific objects in the different types of 3-D model to information of both a spatial and a non-spatial character.

3.7 A POSSIBLE FRAMEWORK FOR 3-D MODELS IN URBAN PLANNING

The different types of model can, for each type, be described as a specific level of abstraction. In the area of spatial planning where 3-D models of physical systems are used, this can vary from photorealistic models to more simple extruded CAD drawings. It is possible to identify at least four different levels of abstraction:

- near reality
- enhanced reality
- enhanced virtuality
- virtuality

All four different kinds of model have specific characteristics.

3.7.1 Near reality

Near reality is the kind of 3-D model in which video or stereoscopic pictures are used to give an illusion of 3-D. These models are mainly image-based (Fig 3.3).

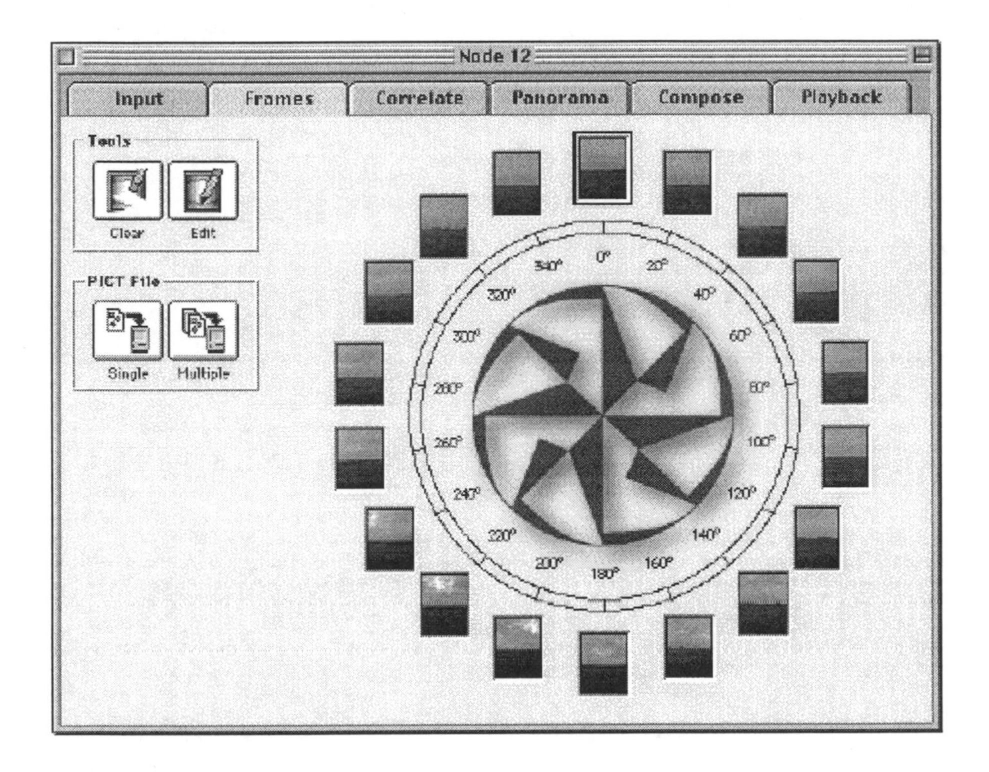

Figure 3.3 Quicktime Virtual Reality (QTVR) is an example of near reality.

3.7.2 Enhanced reality

Enhanced reality uses the video or photographs as a background for new physical objects in the plan. These objects are usually created in and taken from a 3-D CAD model. These new objects can mainly be placed in the foreground of the model. It is more difficult to integrate the objects in the model so that they can be moved behind real objects in the video or in the pictures. This kind of model is used, for example, in feature films such as Jurassic Park and The Lost World. These models are image-based with vectorised features on top.

3.7.3Enhanced virtuality

Enhanced virtuality is, for example, the texture mapping of pictures on to objects in a CAD model. It can also take the form of pieces of digital film that are played on top of a CAD model. The idea is that the CAD model is the background and the small pieces of photograph or film can give the model a more realistic image. These models are vector-based CADmodels, with images used as texture.

3.7.4Virtuality

Virtuality is a full scale model created without the use of realistic elements. All objects in a virtual model are created as vector-based features in a CAD environment. They can be combined with more abstract patterns stored as images (Figs 3.4 and 3.5).

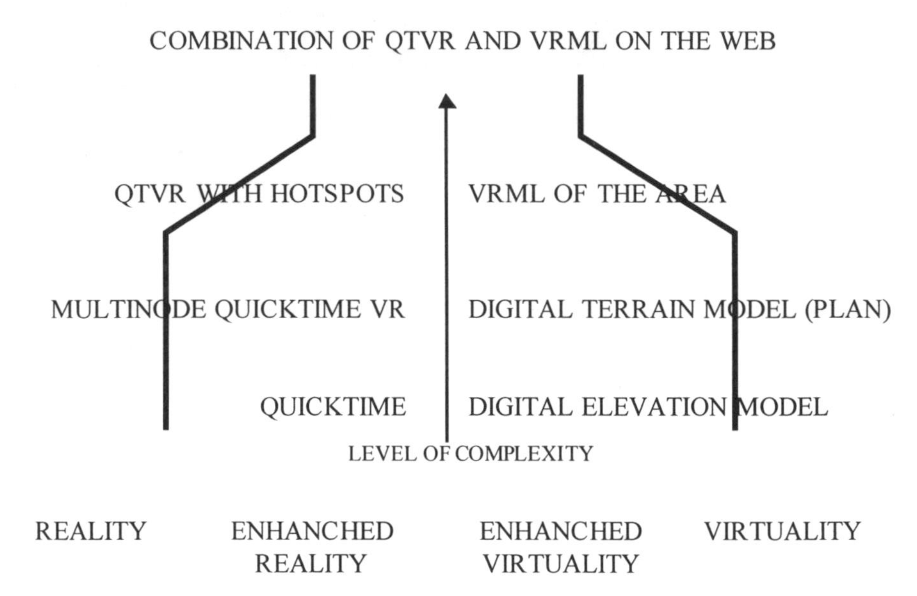

Figure 3.4 A model of how elements from reality and elements created in virtuality can be combined in a hypermedia environment on the Web.

Figure 3.5 A screen dump from an ArcView 3-D scene.

3.8 CONCLUSION

No matter what kind of technology is used in urban planning, the main goal of the planning process is to eliminate misunderstandings in the communication of the planning proposals and to visualise the different alternatives in a planning situation. Communication in the form of a visualisation of the message is the most important issue. To reach a situation in which it will be possible to achieve this goal, it becomes necessary to look at the situation in planning offices today and, especially, to focus on the use of information technology and the need for data, tools for analysis, tools for visualising and tools for communication of the planning documents.

This chapter has shown that there is a need for new ways of representing and communicating geographical information because of the changes that have taken place within geographical information technology and in the handling of spatial referenced data. In particular, within the area of urban planning, there is a need for new tools that can visualise the urban environment from perspectives other than the single one that is available today (Bodum, 1997). The traditional map has its limitations in these surroundings, because of the complex data structures in the urban areas and because it does not suit the computer monitor as well as it suits a printed edition.

There is no doubt that networking and hypermedia will take over, and that this will make the ideal background for a real breakthrough of the hypermap. This concept is also known for its object-oriented approach to spatial data handling. There will be a change from seeing the GIS as one system on one computer to a much more dynamic structure, in which both the data and the handling techniques can be distributed via the Web. In this transitional phase, it is necessary to look at the map as the carrier of information, because this term is constantly used in the GIS community and among urban planners as the primary visualisation of a part of a geographical database. This concept of a solitary medium of communication for geographical information will be exchanged with new methods for combining two or more different types of media in the description of geographical space.

REFERENCES

APPLE COMPUTER, 1996, QuickTime VR—an overview of Apple's QuickTime Technology, http://qtvr.quicktime.apple.com.

BATTY, M., 1997a, The computable city, *International Planning Studies*, **2**, 155–73.

BATTY, M., 1997b, Digital planning: preparing for a fully wired world, in SIKDAR, P. K., DHINGRA, S. L. and KRISHNA RAO, K. V. (Eds.), *Computers in Urban Planning and Urban Management*, New Delhi: Narosa.

BATTY, M., 1997c, Virtual geography, *Futures*, **29**, 337–52.

BODUM, L., 1993, Digital film som medie i GIS, *Landinspektøren*, **36**(4–93), 498–503.

BODUM, L., 1994, Hypermedia-aided GIS in urban planning, Skriftserien 147, Department of Development and Planning, Aalborg University.

BODUM, L., 1995, World Wide Web og geografiske informationer, *Landinspektøren*, **37**(4–95), 466–75.

BODUM, L., 1997, Towards hypermedia-aided GIS in local planning, in Craglia and Couclelis (1997), pp. 544–55.

CAPORAL, J. and VIÉMONT, Y., 1997, Maps as a metaphor in a geographical hypermedia system, *Journal of Visual Languages and Computing*, **8**(1), 3–25.

COTTON, B. and OLIVER, R., 1993, *Understanding Hypermedia*, London: Phaidon.

CRAGLIA, M. and COUCLELIS, H. (Eds.), 1997, *Geographic Information Research ♁Bridging the Atlantic*, London: Taylor & Francis.

CRAGLIA, M. and RAPER, J., 1995, GIS and multimedia, *Environment and Planning B: Planning and Design*, **22**(6), 634–7.

DANAHY, J. W. and WRIGHT, R., 1988, Exploring design through 3-dimensional simulations, *Landscape Architecture*, **78**(5), 64–71.

DRIESSEN, R. J., 1993, Geographic information space, *URISA Journal*, **5**(2), 68–72.

DRUCKER, D. L. and MURIE, M. D., 1992, *Quicktime Handbook*, Carmel: Hayden.

ERICKSON, T., 1993, From interface to interplace: the spatial environment as a medium for interaction, in FRANK, A. U. and CAMPARI, I. (Eds.), *Spatial Information Theory, a Theoretical Basis for GIS,* Lectures in Computer Science,

Vol. 716, Heidelberg, Springer-Verlag, pp. 391–405.

FISHER, P. (Ed.), 1995, *Innovations in GIS 2*, London: Taylor & Francis.

FISHER, P., 1997, Concepts and paradigms of spatial data, in Craglia and Couclelis (1997), pp. 297–307.

FONSECA, A. and CÂMARA, A., 1997, The use of multimedia spatial data handling in environmental impact assessment, in Craglia and Couclelis (1997), pp. 556–70.

GIERTSEN, C., *et al.*, 1997, An open system for 3D visualisation and animation of geographic information, in Craglia and Couclelis (1997), pp. 571–87.

JENSEN, M. T., *et al.*, 1993, Detaljering af digitale 3D–bymodeller, *Landinspektøren*, **36**(4–93), 484–8.

JOHNSTON, K. M., 1990, Geoprocessing and geographic information system hardware and software: looking towards the 1990s, in Scholten and Stillwell (1990), pp. 215–27.

JONES, R. M., *et al.*, 1994, An analysis of media integration for spatial planning environments, *Environment and Planning B: Planning and Design*, **21**(1), 121–33.

KIIB, H. and VEIRUM, N. E., 1993, Hypermaps in urban planning, Skriftserien 109, Department of Development and Planning, Aalborg University, Denmark.

KOOP, O., 1995, Reality and realities: a brief flight through the artificial landscapes of the virtual worlds, in ORMELING, F. J., *et al.* (Eds.), *Teaching Animated Cartography*, Madrid: International Cartographic Association.

KRAAK, M.-J., 1995, The map beyond geographical information systems, in Fisher (1995), pp. 163–8.

LARSEN, T. K., *et al.*, 1998, IT phone home, Unpublished paper, Department of Development and Planning, Aalborg University, Denmark.

LAUREL, B. (Ed.), 1990, *The Art of Human-Computer Interface Design*, Reading, MA: Addison-Wesley.

LAURINI, R. and THOMPSON, D., 1992, *Fundamentals of Spatial Information Systems*, London: Academic Press.

LIGGETT, R. S. and JEPSON, W. H., 1993, An integrated environment for urban simulation, in *Third International Conference on Computers in Urban Planning and Urban Management*, Atlanta, GA.

MAGUIRE, D. J., 1989, *Computers in Geography*, Harlow: Longman Scientific & Technical.

MAGUIRE, D. J., *et al.*, 1991, *Geographical information Systems*, London: Longman.

MARBLE, D. F., 1990, The potential methodological impact of geographic information systems on the social sciences', in ALLEN, K. M. S., GREEN, S. W. and ZUBROW, E. B. W. (Eds.), *Interpreting Space: GIS and Archaeology,* London, Taylor & Francis.

MITCHELL, W. J., 1995, *City of Bits*, Boston, MA: The MIT Press.

NEGROPONTE, N., 1995, *Being Digital*, New York: Alfred A. Knopf.

NIELSEN, J., 1993, *Hypertext & Hypermedia*, London: Academic Press.

PARSONS, E., 1994, Visualisation techniques for qualitative spatial information, in ARRAGON, P. (Ed.), *EGIS/MARI '94*, Paris: EGIS Foundation.

RAPER, J., 1997, Progress towards spatial multimedia, in Craglia and Couclelis (1997), pp. 525–43.

RAPER, J. and LIVINGSTONE, D., 1995, The development of a spatial data explorer for an environmental hyperdocument, *Environment and Planning B: Planning and Design*, **22**(6), 679–88.

SCHOLTEN, H. J. and STILLWELL, J. C. H. (Eds.), 1990, *Geographical Information Systems for Urban and Regional Planning*, The GeoJournal Library, Dordrecht: Kluwer Academic.

SHIFFER, M. J., 1992, Towards a collaborative planning system, *Environment and Planning B: Planning and Design*, **19**(6), 709–22.

SHIFFER, M. J., 1995, Interactive multimedia planning support: moving from stand-alone systems to the World Wide Web, *Environment and Planning B: Planning and Design*, **22**(6), 649–64.

TOMLINSON, R. F., 1988, The impact of the transition from analogue to digital cartographic representation, *The American Cartographer*, **15**(3), 249–61.

TUFTE, E. R., 1990, *Envisioning Information*, Cheshire, England: Graphics Press.

VEIRUM, N. E., 1992, Hyperkort (og godt) om GIS og multimedier, *Landinspektøren*, **36**(1–92), 35–43.

WIGGINS, L. L. and SHIFFER, M. J., 1990, Planning with hypermedia: combining text, graphics, sound and video, *Journal of the American Planning Association*, **56**(2), 226–35.

CHAPTER FOUR

Augmenting transportation-related environmental review activities using distributed multimedia

Michael J. Shiffer

Planning Support Systems Group, Department of Urban Studies & Planning, Massachusetts Institute of Technology, 77 Massachusetts Avenue, Room 9-514, Cambridge, MA 02139, USA

4.1 INTRODUCTION

This chapter explores the development of spatial multimedia representational aids to support various transportation-related environmental review activities. The implementations provide a method of interacting with planning analysis tools using graphical interfaces that are implemented through World Wide Web (WWW) and stand-alone multimedia systems for group decision support. These applications of technology have the capacity to better describe abstract representations of transportation-related environmental impacts (such as noise, air quality and congestion) through the juxtaposition of maps, video, text, sound and other media.

This chapter is intended to provide an overview of how spatial multimedia has been applied to a problem related to representations of automobile traffic. Given the ongoing nature of this research, this chapter will simply describe a problem context and a given method of intervention. It is not intended to be a review of the body of work in this area; nor does it contain conclusive results about the effect of these tools on collaborative group processes. It will close with an identification of future directions for this research.

4.2 PROBLEM CONTEXT

When one considers the impact of significant urban improvements (in the United States and elsewhere), the work of planners, architects and engineers is frequently reviewed in public contexts such as city planning meetings. Thus, it is important for planning-related professionals to effectively communicate the details of existing problems and their proposed solutions to groups of people with vastly different levels of knowledge. Recent trends (and several legislative mandates[1]) have led to an evolution from the 'passive presentation' of a proposed plan, to an 'active dialog' in which alternative scenarios may be generated and explored by decision-makers who draw upon the expertise of various parties.

Planning reviews provide an example of such an 'active dialog'. These situations frequently include recollections about the past, descriptions of the present and speculation about the future of a given area. In recollecting the past history of a site, conversations may revolve around what was said, what was done or what a place was like (among other things). For example, members of a group may try to recall the impact of past developments to better understand what may lie ahead, given similar circumstances. Where the recollection is about fairly structured recent activities (such as past planning meetings), the conversations can be supported with records and systematic documentation of past interactions. However, access to this information is frequently difficult and can be dependent on a specialized information recording and retrieval strategy (such as the systematic use of a secretary or recording mechanism). Furthermore, such methods of recollection rarely incorporate spatial referencing using historical maps or site plans.

Where systematic documentation is lacking, the high degree of dependence on human memory can lead to problems based on the inconsistency of individual memories. This is exacerbated by the fact that personal points of view tend to be subjective. For example, someone may recall traffic on a particular street to have been heavy, while another may think of the same stretch of roadway as lightly traveled. Where there is a lack of documentation or data to support these recollections, arguments related to inconsistent memories are likely to persist. These arguments can dominate a discussion and shift the focus of a meeting from the matters at hand.

Descriptions of present conditions are generally used to familiarize meeting participants with an area being discussed so that everyone can work from a common base of knowledge. These descriptions frequently involve some sort of spatial referencing. While many of these references are verbal (i.e. 'by the river' or 'on the site of the former railway station'), such references become increasingly inappropriate as the level of familiarity with the site on the part of meeting participants is lacking. For example, the term 'on the site of the former railway station' is completely meaningless to meeting participants who are unfamiliar with the area being discussed.

Finally, speculation about the future of an area frequently involves the application of past experiences to the future using informal mental models. More recently, formalized mechanisms for such speculation have been made available through the use of computer-based analysis tools for prediction and modeling (cf. Klosterman *et al.*, 1992). In public review contexts, however, such speculation has traditionally been handicapped by a lack of immediate response from the available analytic tools. Furthermore, many of these tools lack the descriptive abilities of

[1]One such mandate in the United States is the Intermodal Surface Transportation Efficiency Act of 1991 (ISTEA). Among other things, ISTEA calls for a greater degree of public review of proposed transportation improvements (U.S. Dept. of Transportation, 1992).

images and human gestures. For example, while an analytic tool may provide a quantitative measure of the estimated automobile traffic level at a given intersection, it is not able to provide an example of 'how crowded the streets will be' to individuals who are not technically sophisticated (Shiffer, 1995). Similarly, quantitative measurements that describe environmental impacts, such as noise, may be meaningless to the lay person. This could serve to cloud the public's understanding of potentially beneficial projects and contribute to misunderstandings (Forester, 1982).

4.3 METHOD OF INTERVENTION

Many of the information usage difficulties described above can be addressed through a creative application of multimedia information technologies (cf. Laurini and Milleret-Raffort, 1990; Wiggins and Shiffer, 1990; Câmara *et al.*, 1991; Fonseca *et al.*, 1993; Polyorides, 1993). More recently, MIT's Department of Urban Studies and Planning has been working with several government agencies and private firms to identify effective combinations of tools and techniques for the support of city planning meetings. The overall goal of this research is to improve the communication of planning-related information with a specific focus on the environmental effects of proposed urban interventions. The intention is to close the gap between planners and their constituents. This can be accomplished through (1) the identification of specific gaps in the communication of planning-related information, (2) the augmentation of typically abstract environmental representations using direct manipulation interfaces and multimedia representational aids, and (3) the effective delivery and evaluation of these mechanisms.

The intended result is to make analytic tools and their outputs, more manipulable, understandable and appealing, so that information that would normally be meaningless and intimidating to the lay person can be more effectively comprehended. As part of this ongoing research, a prototype Collaborative Planning System (CPS) has been developed to support planning activities in several cities (Shiffer, 1995a).

The CPS has been constructed with a hardware infrastructure consisting of a personal computer (such as an Apple Macintosh or an IBM-compatible PC running the latest version of Microsoft Windows), an audio/video digitizing card, a flatbed scanner, a portable video camera and plenty of digital storage space. The CPS also relies on: digital video and graphics editing software (such as Adobe Premiere and Photoshop) for the capture and manipulation of images; thematic mapping packages (such as MapInfo or ESRI's ArcView) for the generation of maps; and multimedia authoring tools (such as Allegiant's Supercard or Macromedia's Director) for data integration and multimedia programming. Recent work has also led to the use of authoring tools that can support access to information via the World Wide Web (WWW).

The CPS uses a multimedia information base that is projected on the wall of a meeting room. The shared display tends to be more conducive to conversation, less intimidating to novices and less costly than individual computer monitors. Participants interact with the system using cordless pointing devices or a technical facilitator to elicit information about selected geographic areas through a direct manipulation graphical interface. By allowing the user to manipulate graphics, the system's interface translates the users' intentions into commands that the computer can understand (Norman, 1986). The computer's output is translated into concepts that the human can understand through the use of multimedia representational aids such as sound and images.

Various planning-related topics such as automobile traffic scenarios, neighborhood densities, building heights and zoning restrictions can be explored. In addition to this, the impact of proposed facilities can be visualized using a video sketching tool that superimposes proposed representations onto motion video clips of existing sites. Finally, maps and other information contained in the system can be annotated with text, audio or video. The annotation component creates an object, such as a polygon or icon, that can 'store' video clips, sounds and users' comments. Verbal comments that are made during the meeting, and gathered in this manner, can later be transcribed and incorporated into a text field that can facilitate key word searching.

The development of the prototype CPS has led to the identification of a broad set of issues ranging from institutionalization to technical infrastructure (some of which are discussed in Shiffer, 1995b). The remainder of this chapter will focus on the issue of representation by looking more closely at how we can represent automobile traffic scenarios using spatial multimedia.

4.4 A CHALLENGE: REPRESENTING AUTOMOBILE TRAFFIC SCENARIOS

Many potential urban developments are evaluated for their impact on a community's existing transportation infrastructure. In particular, this impact is frequently represented as a projected change in traffic conditions. Several measures have been used to represent perceived changes in traffic conditions. Some of the more popular measures include average daily traffic (ADT) measures and standardized levels of service (LOS).

ADT is usually expressed as a numeric value that describes the average number of vehicles passing a fixed point over a 24 hour period. In the context of a city planning meeting, one use of ADT might be to describe the impact of a new development on an existing traffic network. For example, a traffic specialist might convey this information to a city planning meeting by saying 'if the new shopping center is built, the ADT at that location is projected to rise from 11 300 to 14 500'. Unfortunately, this numeric value may have very little meaning for meeting participants who are not familiar with these measures. When someone in the meeting asks for clarification, the traffic specialist might try to relate the value of 14 500 to another street by saying 'if the new shopping center is built, the traffic on Main Street will become more like the traffic on State Street'.

While this could serve to clear up some internal confusion, it is highly dependent on participants having similar internal representations of the traffic on both State and Main Streets under comparable conditions. This could prove to be problematic, as individual perceptions may differ based on the conditions under which these perceptions have been formed. For example, if one person's sole perception of Main Street has been based on exposure to the street during rush hour, it is likely to be significantly different than another person's, which may have been formed while driving on the same street in the mid-morning. Furthermore, people who have no concept of Main Street will have a great deal of difficulty understanding the analogy that has been drawn by the traffic specialist.

4.5 EARLY ATTEMPTS

An early attempt to better represent abstract measures of automobile traffic was undertaken in the development of the CPS prototype described above and more specifically described by Shiffer (1995a). The effort to convey ADT involved the juxtaposition of several traffic representational modes, including numeric values, a fluctuating graphical bar and digitized video segments (see Figure 4.1). The graphical bar and numeric value would fluctuate to indicate varying levels of traffic as the user(s) would move a pointer around an on-screen map. The use of a map as an interface mechanism in this context would provide a rich example for those meeting participants who were familiar with the behavior of the streets in the area.

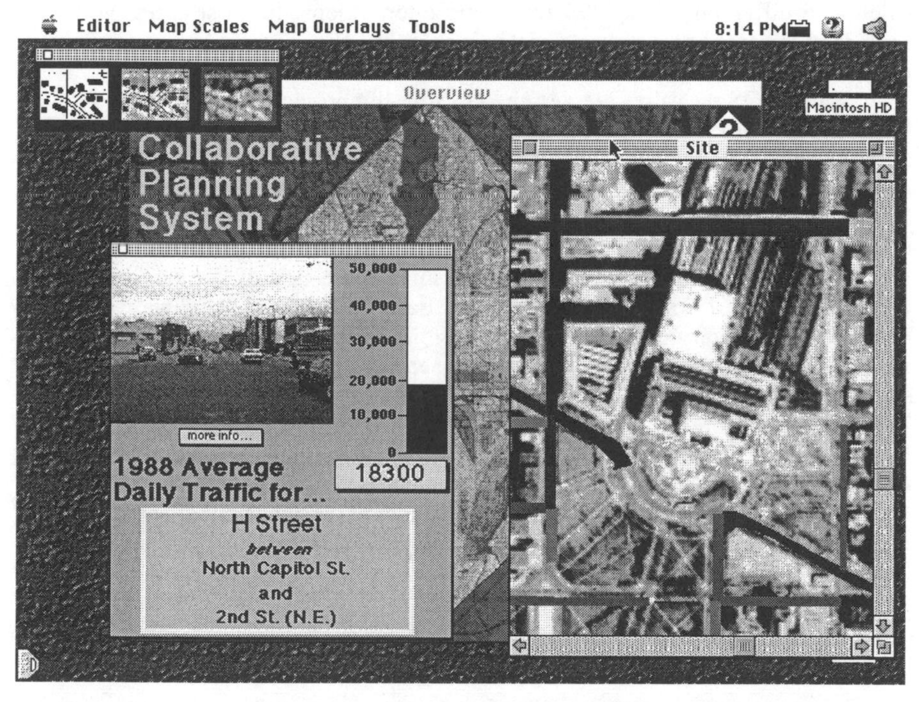

Figure 4.1 Average Daily Traffic (ADT) representations as implemented in a prototype Collaborative Planning System (CPS).

It was theorized that a clearer description of traffic conditions could be provided through the introduction of motion video clips. This mode of representation could provide more concrete examples of how a given environment is affected by changes in traffic levels. However, such representations may only be valid if they have been drawn from a set of comparable examples. (As in the traffic specialist's analogy, a digitized video example of traffic conditions taken at 10:30 hours on a Tuesday morning is not likely to be comparable to a sample of traffic at the same location taken during a Friday evening rush hour). Such representations may be valid, however, if they are drawn from a library of comparable examples where the conditions under which the samples were taken are made explicit.

Nevertheless, an image of traffic at a given time only provides a partial view of proposed conditions (even when juxtaposed with other representations). Since traffic at a fixed location is likely to change, we are faced with the challenge of capturing the behavioral character of the traffic segment.

4.6 REPRESENTING LEVELS OF SERVICE

One approach, as described in the Highway Capacity Manual (Transportation Research Board, 1985) has been to classify or 'grade' traffic conditions using a measure known as Level of Service (LOS). LOS measures differ by the environment that they seek to describe. For example, LOS for continuous flow traffic is calculated differently than LOS for traffic intersections. In most cases, however, LOS is represented as a discrete scale from 'A' to 'F', where 'A' represents very good traffic conditions and 'F' represents very poor traffic conditions.

While LOS measures are designed to simplify projected traffic levels through the assignment of a letter grade, this simplification continues to provide a level of abstraction that is frequently difficult for members of the general public and decision-makers to comprehend. This is not so much due to complexity as it is due to a lack of rich descriptions of conditions that the various participants in a meeting can effectively relate to.

An early attempt to make LOS measures more descriptive has involved a juxtaposition of LOS representations as described in the Highway Capacity Manual with a library of generic indicative images. The images are taken as digital video clips that represent approximately 10 seconds of traffic at a given LOS. After selecting a traffic situation, such as three lanes of continuous flow traffic (as is illustrated in Figure 4.2), the user can then explore how that traffic is likely to behave under various predicted LOS conditions. The interaction occurs when the user selects an LOS which yields the associated digital video clip and a brief textual description of traffic and driver behaviors under those conditions. The user can then select other LOS scenarios for direct or pair-wise comparisons of traffic behaviors under several sets of conditions.

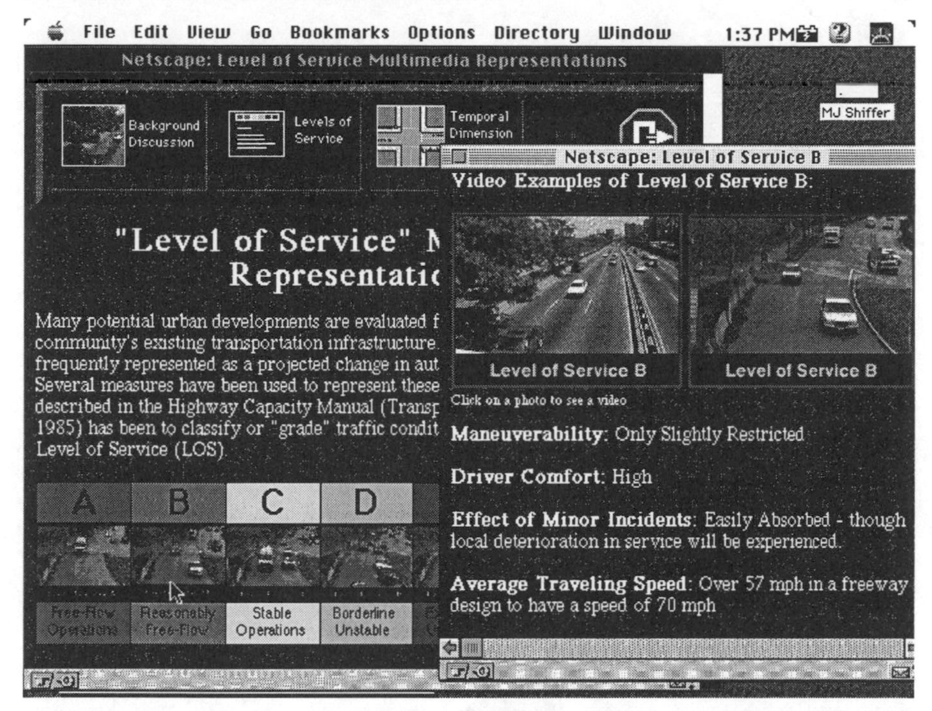

Figure 4.2 Multiple representations of Level of Service (LOS) data for a continuous flow highway as represented on the WWW.

This electronic library of LOS descriptions can be effectively delivered to various planning organizations using the WWW, provided that the representations are drawn from a fairly generic set of situations. This data can be provided and maintained in a distributed manner by these organizations using the overall structure that is being developed in this research as a guideline. While this type of analysis can lead to a broadened understanding of how one LOS for a given set of physical conditions compares to another, it assumes that the user has some concept of existing or predicted LOS for their area. Where this is not the case, it may be desirable to predict future, or describe existing, LOS using the standard metrics published in the Highway Capacity Manual. Future implementations will likely include a mechanism for calculating and predicting LOS.

4.7 CAPTURING THE TEMPORAL DIMENSION OF TRAFFIC

The representation that is characterized by LOS may be useful for describing a given set of traffic conditions. However, it is incomplete in that it does not effectively convey the temporal aspect of how the traffic will behave over a given period of time. To capture this additional aspect of automobile traffic representations, we have begun to experiment with time-lapse photography as a mode of representation.

While time-lapse photography in an urban setting is not a new technique (cf.

Whyte, 1980) it presents some special problems for capturing images of an urban environment in a systematic manner. One problem is that of camera placement. In a context in which the camera needs to capture traffic conditions, it is frequently desirable to view the street or intersection from a fixed aerial perspective. This issue has been addressed in some communities by the placement of cameras on top of traffic signals. However, this type of implementation is costly, as it typically requires several cameras in weather-proof housing, with a central control center where taping can take place. A more economical approach can be undertaken by identifying several key buildings that have favorable views of the traffic conditions that need to be recorded. Aside from the significant organizational obstacles that one may run into when trying to gain access to one of these buildings, there is the issue that such access (when available) is likely to be inconsistent at best and subject to various obstacles that may restrict the field of view.

Our approach has been to use a lighting stand that is extendible to 6 meters in height to support a small consumer-grade camcorder. The lighting stand can be deployed either at street level (on a sidewalk) or from within a vehicle with an operable sunroof. While the use of a vehicle provides a measure of safety (and a high-capacity power source for the video camera in the form of the vehicle's cigarette lighter), deployment on a sidewalk offers a greater degree of flexibility.

Another consideration has been whether to record tape as continuous time-lapse, continuous real-time or regular interval sampling throughout the day. Taping continuous time-lapse typically requires special video equipment that allows for such recording.[2] It also requires that some attention be paid to the camera's placement and its relative safety. Taping continuous real-time affords a bit more flexibility in that it can be accomplished using consumer-grade video equipment. It also provides the opportunity to vary video sampling rates (measured in frames per second or minute) by compressing time digitally rather than doing so in the field. Such time compression can occur either during digitization or by re-sampling video that has already been digitized by using a computer-based digital video editing application such as Adobe Premiere. Nonetheless, this method tends to be quite labor-intensive, as it requires frequent changing of tapes and batteries in the field and a significant amount of time in the digitizing process.

An approach that we have found to be flexible and less labor-intensive has been to sample real-time video at regular intervals throughout the day. The intervals can be determined by the perceived variability of traffic at that location. For instance, during peak hours when traffic is highly variable, one would sample at closer intervals than during midday when traffic conditions are relatively stable.

In the example illustrated by Figure 4.3, an intersection was observed for an entire (90 second) light cycle every 90 minutes. By using masking tape on the pavement to mark the tripod's location, we were able to remove the equipment between iterations, thus eliminating the need for constant monitoring. The resulting video clips were digitized at one frame per second for detailed analysis. All of the video clips were complied into a (less detailed) overview that played through the entire day's sequence in 15 seconds.

The resulting movie was enough to give a general indication of traffic behav-

---

[2]Some commercially available consumer camcorders have a built-in interval recording feature that allows for quasi-time lapse, where a length of video definable by the user is recorded after a set duration (also definable by the user).

ior at a specific location over the course of an entire day. While this could effectively augment an ADT value, this representation is not very useful by itself. However, when other data (such as a dynamic plot of traffic volume) is added, it becomes much more descriptive. Furthermore, when it is compared to representations of traffic at other locations, it has the capacity to provide a much greater degree of understanding for both lay people and traffic specialists.

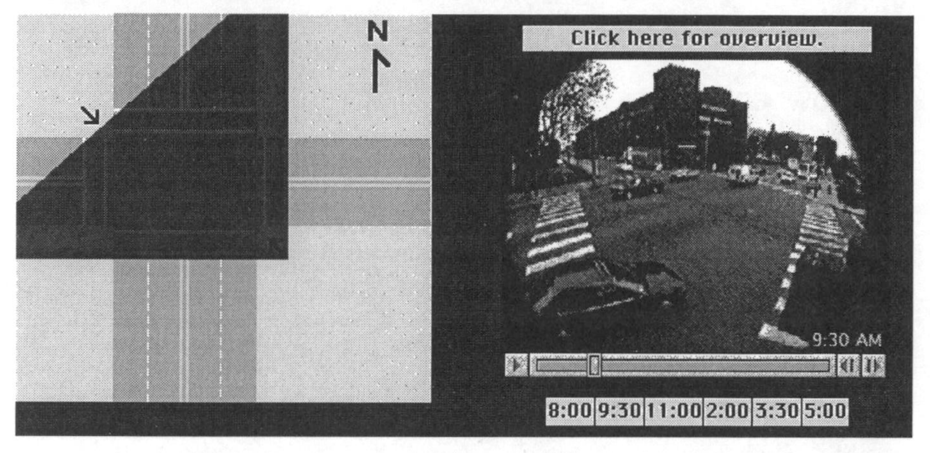

Figure 4.3 An example of real-time video that was sampled at regular intervals throughout a day and then digitally compressed into a time-lapse representation.

4.8 CONTINUING ISSUES

The goal of this implementation is to provide several examples of how multimedia representations of standard automobile traffic measures can be effectively delivered to support planning activities. Furthermore, an intention of this aspect of the project is to provide a set of procedures that local governments can follow in the creation and delivery of a standardized set of libraries that can be drawn upon in various planning contexts. The wide range of accessibility that is afforded by the distribution of this information via the WWW can result in a more effective public review of the potential effects of transportation planning-related projects. This can allow many smaller to medium-sized communities to benefit from a large information base constructed to reflect the experiences of others with similar problems and challenges. Thus, many communities without the resources necessary to perform exhaustive analyses can learn from the experiences of others.

More work will need to be done to attain a better level of understanding of the processes in which transportation specialists represent traffic. Furthermore, we need to adapt existing metrics (or develop new ones if necessary) to assess the relative impact of these representational aids. We also need to continue to refine our models of delivery for these tools. While the WWW holds tremendous promise of distributed multimedia libraries such as the one described here, issues of speed, reliability and interactivity with forecasting models will need to be addressed if this technology is going to play a meaningful role in interactions between transportation planners and the public.

Through an extended application of this research, we will be able to move beyond a simple implementation of new technologies to a sustained period of implementation, in which the impacts of this technology could be studied systematically over an extended period. This will result in a better understanding of the technology's role in shaping planning processes and their effects on planning related discussions. The intended results of these efforts are more meaningful explorations and exchanges about the future of urban areas.

ACKNOWLEDGMENTS

Partial support for this project was provided by the US Department of Transportation, Bureau of Transportation Statistics, and the MIT Center for Transportation Studies Contract #DTRS–57–92–C–00054 #45. John G. Allen, Joey B. Ferreira, Christina Gouveia, Albert Wong and Roberto Ordorica have all contributed significantly to this research.

REFERENCES

CÂMARA, A., GOMES, A. L., FONSECA, A. and LUCENA E VALE, M. J., 1991, Hypersnige, a navigation system for geographic information, in *Proceedings of the European GIS Conference*, Brussels, Belgium, April, pp. 175–9.

FEDERAL TRANSIT ADMINISTRATION, 1992, Intermodal Surface Transportation Efficiency Act of 1991: flexible funding opportunities for transit, moving America into the 21st century, Washington, DC: US Department of Transportation.

FONSECA, A., GOUVEIA, C., FERREIRA, F. C., RAPER, J. and CÂMARA, A., 1993, Adding video and sound into GIS, in *EGIS'93 Conference Proceedings*, Genoa, pp. 176–87.

FORESTER, J., 1982, Planning in the face of power, *Journal of the American Planning Association*, **48**, 67–80.

KLOSTERMAN, R. E., BRAIL, R. K. and BOSSARD, E. G., 1992, Spreadsheet models for urban and regional analysis, New Brunswick, NJ: Center For Urban Policy Research.

LAURINI, R. and MILLERET-RAFFORT, F., 1990, Principles of geomatic hypermaps, in *Proceedings of the 4th International Symposium on Spatial Data Handling*, Zurich, pp. 642–51.

NORMAN, D. A., 1986, Cognitive engineering, in NORMAN, D. A. and DRAPER, S. W. (Eds.), *User Centered System Design: New Perspectives on Human-Computer Interaction*, Hillsdale, NJ: Lawrence Erlbaum, pp. 31-61..

POLYORIDES, N., 1993, An experiment in multimedia GIS, in *Proceedings of the European GIS Conference*. Genoa, Italy, April, pp. 203–12.

SHIFFER, M. J., 1992, Towards a collaborative planning system, *Environment and Planning B: Planning and Design*, **19**, 709–22.

SHIFFER, M. J., 1995a, Multimedia representational aids in urban planning support systems, in MARCHESE, F. (Ed.), *Understanding Images*, New York:

Springer-Verlag, pp. 77–90.

SHIFFER, M. J., 1995b, Environmental review with hypermedia systems, *Environment and Planning B: Planning and Design*, **22**, 359–72.

TRANSPORTATION RESEARCH BOARD, NATIONAL RESEARCH COUNCIL, 1985, in *Highway Capacity Manual*, Washington, DC: Transportation Research Board, National Research Council.

WHYTE, W. H., 1980, The social life of small urban spaces, Washington, DC: Conservation Foundation.

WIGGINS, L. L. and SHIFFER, M. J., 1990, Planning with hypermedia: combining text, graphics, sound, and video, *Journal of the American Planning Association*, Spring, 226–35.

CHAPTER FIVE

Digital video applied to air pollution emission monitoring and modelling

Francisco C. Ferreira
Environmental Systems Analysis Group, New University of Lisbon, Quinta da Torre, P-2825 Monte de Caparica, Portugal

5.1 INTRODUCTION

Recent advances in multimedia technologies allow the capture and storage of video images with relatively inexpensive computers. Specialized image processing software enables the processing of the images and a detailed analysis of their features. As an expansion of the standard use of multimedia capabilities, video images can be used for remote sensing purposes.

In environmental science, video is usually used as a substitute for acrial photography and for some specific applications involving different wavelength sensors.

In air pollution, the development of models and monitoring strategies for single sources such as industrial stacks is now very comprehensive. However, these models, which are used even for regulatory purposes, depend on high-quality meteorological data, which may be difficult to obtain.

Multimedia technologies can overcome part of the problem of gathering a model's input data, and enabling a model's calibration and validation, and can also be used to present images of simulated situations. Therefore, the application of multimedia as an additional tool to understand the dispersion of air pollutants may be very valuable. By giving better quality results, it may help to predict potential polluted areas and to design monitoring networks.

5.2 BACKGROUND

5.2.1 Multimedia and videography

The new possibilities offered by personal computers with regard to multimedia processing have made fast and improved image digitizing and processing capabilities much more accessible.

Videography is the process of capturing, processing and understanding video images (Vlcek, 1988). For certain problems, digital video interactive processing can be used to represent space and time more adequately than traditional numerical approaches (Wodtke, 1993). The use of video images is particularly important in remote sensing (Campbell, 1987; Lillesand and Keifer, 1987). Some of the advantages of video with traditional remote sensing techniques are the availability of the images in real time for processing, and the high sensitivity of video cameras (Everitt, 1988; King, 1991; Mausel *et al.*, 1992).

Some applications can be found that use video technology for monitoring purposes: tracking people in streets to count them and to evaluate their degree of movement (Rourke and Bell, 1992); vehicle counting and identification of vehicle type (Michalopoulos and Wolf, 1990; Kilger, 1992); vehicle emissions measurements using infrared cameras (Lawson, 1990; Stephens and Cadle, 1991; Zhang, 1993); and the detection of chemical clouds either by infrared (Althouse and Chang, 1991) or by ultraviolet for sulphur dioxide (McElhoe and Conner, 1986).

5.2.2 Image processing and cameras model

In order to perform the calculation of the distances that exist on a real image based on a captured video image, it is necessary to set up a perspective transformation of planes. Figure 5.1 shows the model for this transformation.

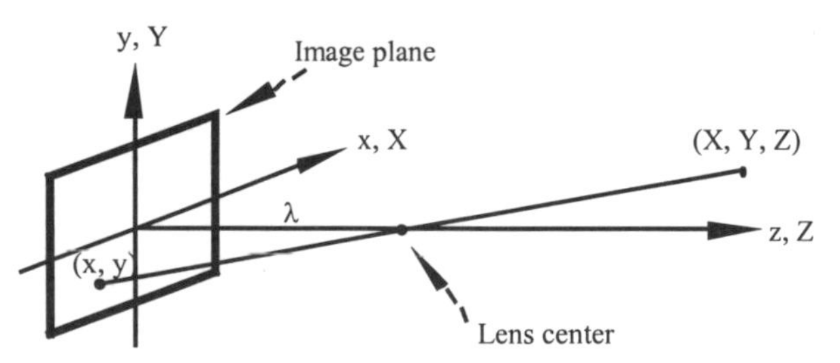

Figure 5.1 The basic model of the imaging transformation process. The camera co-ordinate system (x,y,z) is aligned with the world co-ordinate system (X,Y,Z) (Gonzalez and Woods, 1992).

λ is the focal length of the lens when the camera is in focus. It is assumed that $Z > \lambda$, which means that all the points of interest are positioned in front of the lens. By using similar triangles it is possible to achieve a relationship between the real point (X,Y,Z) and its projection on the image plane (x,y), as expressed by the following equations:

$$\frac{x}{\lambda} = \frac{X}{Z - \lambda}$$

$$\frac{y}{\lambda} = \frac{Y}{Z - \lambda}$$

Instead of using a 3-D model generated by stereo imaging from two cameras located a few kilometres away from the stack and looking horizontally at the plume, a different approach was chosen. This option was adopted since the quality of the distant camera that looks horizontally at the stack is usually not very high. It is very sensitive to atmospheric conditions (clear versus cloudy sky) and to the contrast created by the sun's position relative to the plume. By calculating the wind direction using a camera that looks up vertically at the gaseous outlet at the stack, then high quality and precise information can be acquired. Therefore, Z and λ are known and X and Y can be determined for each x and y present in the image plane. However, this model assumes that there is no change in the direction of the plume along its pathway, which is not always true.

5.2.3 Air pollution modelling and monitoring

Monitoring stations access local information only. They provide very limited data for spatial air pollution analysis. Therefore, air pollution models are usually applied under various circumstances to evaluate the impact of major pollution sources. Gaussian models are the most widely used, because of their simplicity and the relative accuracy of their results. These models require input data such as the pollutants emission's flow and velocity, the emission's temperature, the air temperature, the wind direction, the wind speed, and atmospheric stability information. The model results are highly dependent on these data, particularly wind direction and atmospheric stability. These parameters are usually assessed by meteorological stations located near to the plant (Zannetti, 1990; Weil *et al.*, 1992; Boubel *et al.*, 1994; Ferreira *et al.*, 1995).

Depending on the type of industry, the emissions flow, the local meteorological conditions and the potential impact over populated areas, flora and fauna, the height of the stack can be more than 150 metres. For most of the time, it is impossible to determine wind direction and wind speed at that altitude, and data from meteorological stations that measure those parameters at a much lower altitude are used. However, wind speed and wind direction measurements can differ considerably between the meteorological station and the stack for different altitudes. This effect is particularly marked under certain atmospheric conditions and at certain locations. To overcome this difficulty, a radio wave system named SODAR can be used to determine wind direction, wind speed and atmospheric stability over a wide range of altitudes. However, this instrumentation is very expensive and its use is not standard.

An air pollution single-source Gaussian model was applied to the described situation to calculate the ground concentrations of the gases. The concentration of a pollutant emitted by a single source at a nearby location is a function of multiple factors, as described in the following equation:

$$c = \frac{Q}{2\pi\sigma_h\sigma_z\left|\bar{u}\right|}\exp\left[-\frac{1}{2}\left(\frac{y_r}{\sigma_y}\right)^2\right]\exp\left[-\frac{1}{2}\left(\frac{h_e - z_r}{\sigma_z}\right)^2\right.$$

where c is the concentration at $r = (x_r, y_r, z_r)$, Q is the emission rate, σ_y and σ_z are the standard deviations (horizontal and vertical) of the plume concentration spatial distribution, u is the horizontal wind speed, and h_e is the effective emission height ($h_e = z_S + \Delta h$, in which Δh is the effective plume rise) (Zannetti, 1990).

5.3 CASE STUDY DESCRIPTION

A demonstration of the developed methodology was applied for a 1000 MW fuel power plant located at Setúbal, Portugal. This plant has four 250 MW generators, with two stacks (each stack for two adjacent groups). Three major pollutants are emitted by the power plant: particulates, sulphur dioxide (SO_2), and nitrogen oxides (NO_x).

A group of seven air quality monitoring stations located within a circle of radius 8 km from the plant take measurements every minute (except for particulates, where the interval is 15 minutes). The data is sent instantaneously via a telephone line to a central computer at the plant.

Two other sources may contribute to the pollution levels found: a cement factory located 20 km west of the power plant, and a pulp-and-paper plant located 4 km southeast. Since north and northwest winds are predominant in the area, it is natural to get some background pollution from other industrial facilities located at some distance in this direction.

Wind direction data was collected from 27 September 1994 to 31 May 1995 at the power plant site. Data was collected by two different instruments: a SODAR, which gathered wind direction and wind speed data at altitudes up to 750 m; and a fixed meteorological station at an altitude of 55 m (30 m above ground). The SODAR data for an altitude of 200 m was selected for the comparison with the meteorological data, since this is the height of the stacks at the power plant (Figure 5.2).

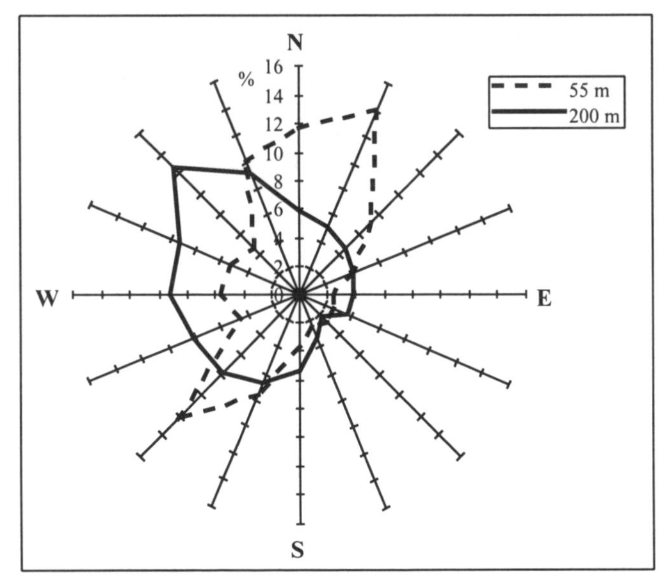

Figure 5.2 A comparison between wind directions measured by a meteorological station at an altitude of 55 m and by a SODAR at an altitude of 200 m.

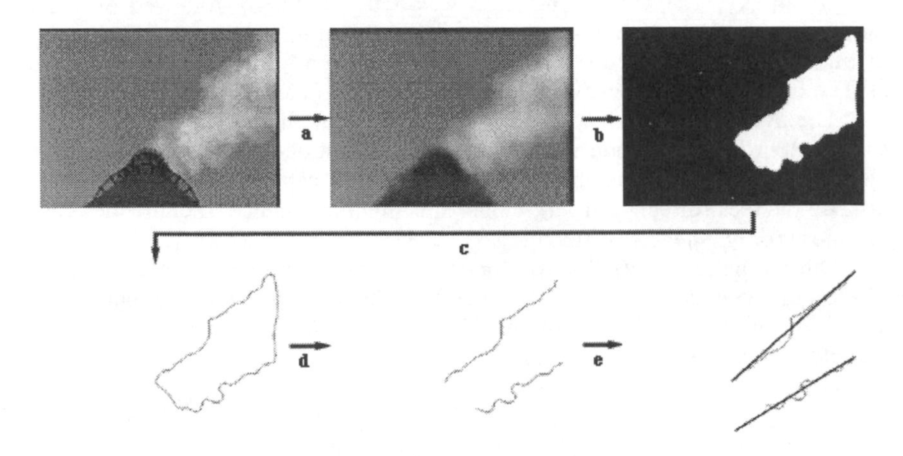

Figure 5.3 Image processing sequential operations for calculation of the wind direction from the vertical camera near the stack.

On the basis of the meteorological data, the predominant winds are from the NNE direction, and also from the NNW and southwest directions. An analysis of the data from the SODAR (at an altitude of 200 m) shows that the pattern is quite different. It is closer to the general circulation characteristics of the area, with predominant winds from the northwest. Therefore, Figure 5.2 shows that, at least in this particular case, the meteorological data would be a great source of error if used for air pollution models or for monitoring evaluation. This is due to the proximity of a large water body and the sea, causing breeze phenomena that may be detected mostly by the meteorological station.

One CCD amateur camera was positioned approximately 2.5 km from the power plant stacks. This camera has an horizontal view of the power plant. Another CCD amateur camera was positioned below one of the stacks, looking vertically at its outlet. This camera covers approximately half of the plume directions. In order to cover all of the possible wind directions, two cameras would be necessary. The images were recorded for further processing. An Apple Macintosh® Centris 650 with a VideoSpigot® video capture board was used to digitize the images collected. Adobe Photoshop® and a computer program developed by the author were used for image processing and calculation purposes.

5.4 RESULTS

The methodology was tested on 11 July 1995, between 19:46 and 20:00 hours (17:46 and 18:00 hours GMT). The sky was relatively clear, with a relative humidity around 70% and an ambient temperature of 20°C.

Figure 5.3 shows the sequence of image processing operations that were performed to extract the direction of plume. The colour image was processed in greyscale. The first filter applied was a gaussian blur (a). This filter would clear some of the detail, but would make more clear the distinction between the plume and the background sky. The second operation was a threshold (b), followed by a trace contour (c). The last operations were to select the contours relative to the directions (d), and to determine the corresponding vectors using a linear regression procedure (e). The final direction is an average of the directions detected from both vectors.

The average wind direction between 19:46 hours and 20:00 hours measured at the meteorological station was 253 degrees (direction from). The measurement calculated with the camera from a group of frames captured during that interval was 262 degrees (direction from). Under the situation studied, the directions were very close (only 9 degrees different).

During the same period of time, a distant camera captured the plume horizontally. Figure 5.4 shows the model that brought together the information from camera A with the wind direction information from camera B, to set up a real co-ordinate system from the image plane.

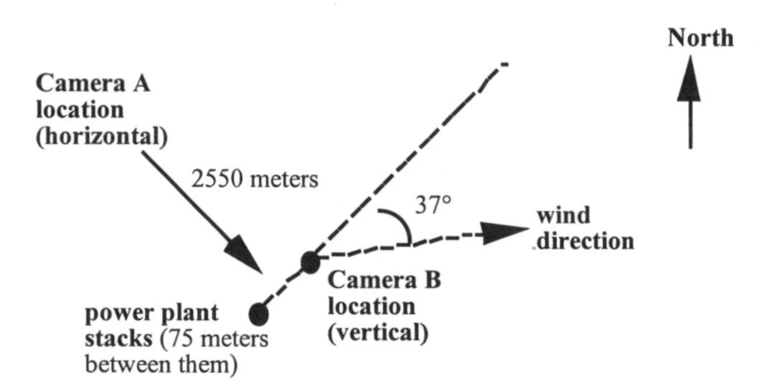

Figure 5.4 The cameras model for image-processing calculations.

Figure 5.5 shows the sequence of operations performed to identify the gaseous plume. The contrast of the captured colour image was enhanced to better reveal the plume. A change detection algorithm was applied to extract the area of the plume (a). By applying the cameras model previously described, the co-ordinate system for the real image was calculated (b). (0,0) are the (X,Y) co-ordinates of the real image.

One of the most important parameters of a Gaussian air pollution model is the effective height of the stack. Through the identification of the plume and using the described cameras model, it is possible to determine the effective height that can be used to validate the model and to improve its performance. Figure 5.6 shows the result for the described situation.

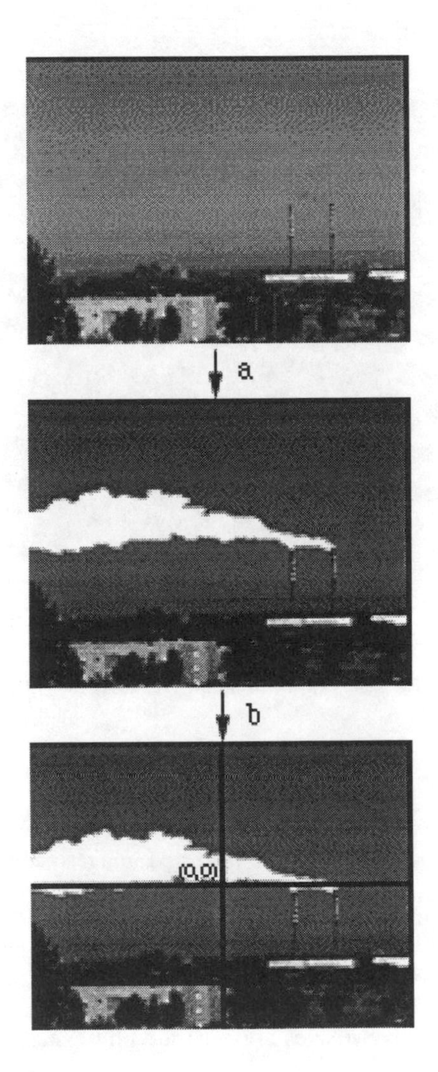

Figure 5.5 The image processing sequential operations for identification of the plume from the horizontal camera far from the power plant.

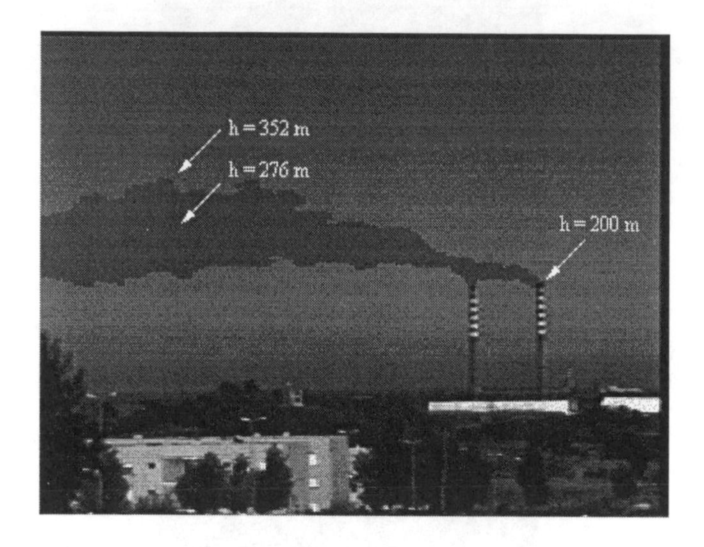

Figure 5.6 The calculated plume rise using the video image.

Another parameter that can be calculated using a video image is wind speed. Figure 5.7 presents a sequence of pre-recorded images captured on 25 June 1996, at 18:33 hours GMT. The average wind speed measurements for that period from the meteorological station were 7.68 m/s and 8.60 m/s, at 10 and 30 m at the tower (altitudes 35 and 55 m), respectively. An average altitude near to the stack can be calculated from the distant camera by applying the model described previously. A pattern within the first image is identified and followed throughout the images sequence. The consecutive images are separated by exactly 6 seconds. A smaller time interval could be used.

The final average wind speed calculated was 11.6 m/s, which is higher than the values measured at the meteorological station. It should be noticed that some error might occur because the plume near to the stack is rising and a fixed height is not normal.

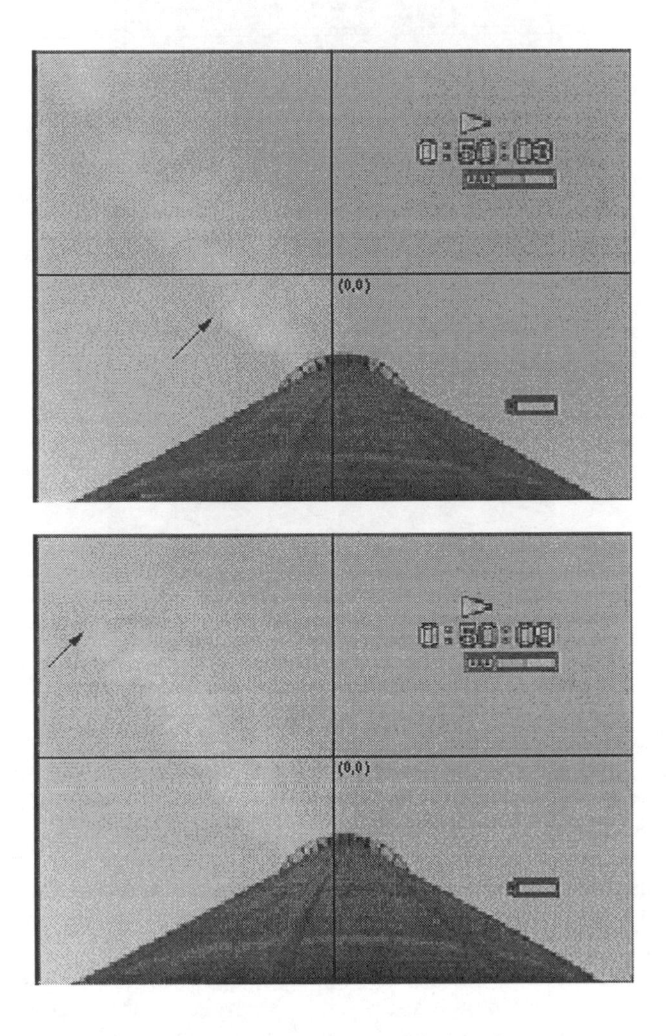

Figure 5.7 The plume pattern followed for 6 seconds.

The stack's outlet gases velocity can also be calculated by using a comparison of two images collected within a certain time interval. From analysing the distance between a pattern identified in both images, it was possible to calculate an outlet velocity of 12.25 m/s, which is slightly higher than half of the expected velocity based on the information after the electrostatic precipitators. This velocity was estimated for a full-production situation. A fraction of this velocity can be calculated for other values of the production power.

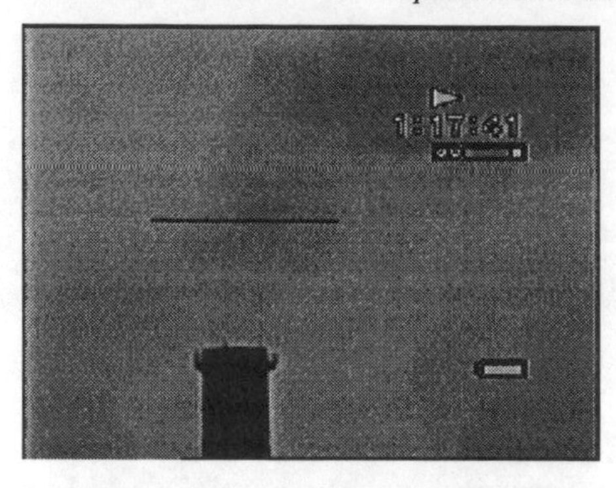

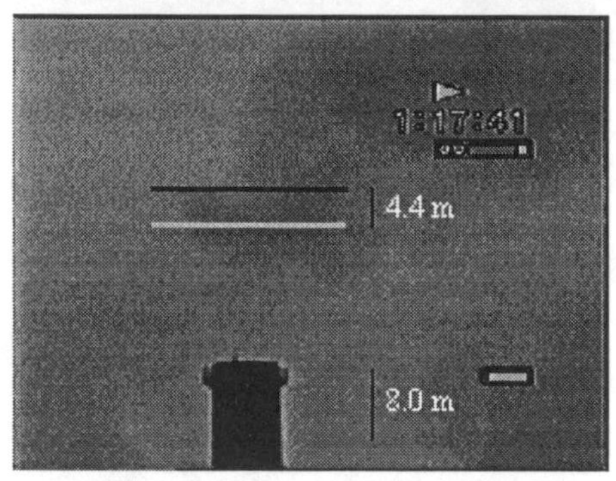

Figure 5.8 The calculation of the velocity of the outlet gases.

A Gaussian air pollution model was run to establish a comparison between the calculated dispersion of the plume from the power plant and its captured image from an horizontal perspective. The emission's parameters were calculated on the basis of sensors installed at each stack, which measure gaseous concentrations, flow, temperature and pressure. The ambient parameters needed for the model were estimated on the basis of the meteorological data collected from the station located at the plant. The wind direction used was the one determined by the vertical camera installed at one of the stacks.

One of the seven air quality monitoring stations was within the pathway of the plume. However, the plume at that site was probably above ground. The predicted concentrations for all of the stability classes considered were zero for both SO_2 and NO_x. The measurements at that station confirm the model estimation, with values less than 1 $\mu g/m^3$.

5.5 CONCLUSIONS

The methodology presented shows that digital video can be an inexpensive way in which to obtain useful air pollution related information. It is possible to track wind direction by analysing the plume at the stack's outlet. Also, the plume's pathway can be identified and the data can be used for air pollution model calibration purposes. Spatial analysis of the data from the video images can help to understand data collected at air quality monitoring stations.

5.6 FURTHER DEVELOPMENTS

Further testing under different atmospheric and lighting conditions will enable the efficiency of the method to be evaluated.

The wind direction may change along the dispersion of the plume. Currently, only the wind direction at the gases outlet is monitored. Therefore, it is assumed that the plume maintains the same direction throughout its dispersion pathway. To overcome this, a stereo imaging system involving two cameras with the same field of view, positioned a few kilometres away, will be tested to track the plume along its path.

ACKNOWLEDGEMENTS

The author would like to acknowledge the data made available and the support for this research by CPPE, Companhia Portuguesa de Produção de Electricidade, SA (Grupo EDP). This work was partially supported by Praxis XXI under research contract Praxis/3/3.2/AMB/04/94.

REFERENCES

ALTHOUSE, M. L. G. and CHANG, C., 1991, Chemical vapor detection with a multispectral thermal imager, *Optical Engineering*, **30**(11), 1725–33.

BOUBEL, R. W., FOX, D. L., TURNER, D. B. and STERN, A. C., 1994, *Fundamentals of Air Pollution*, 3rd Edn, San Diego: Academic Press.

CAMPBELL, J., 1987, *Introduction to Remote Sensing*. New York: The Guilford Press.

EVERITT, J., 1988, Introduction to videography: historial overview, relation to remote sensing, advantages, disadvantages., in *Proceedings of the First Workshop on Videography, American Society of Photogrammetry and Remote Sensing*, Falls Church, VA.

FERREIRA, F., SEIXAS, J. and NUNES, C., 1995, A spatial-based comparison between air pollution modelling and monitoring data, in *Proceedings of the First Joint European Conference and Exhibition on Geographical Information*, The Hague, Netherlands, pp. 448–53.

GONZALEZ, R. C. and WOODS, R. E., 1992, *Digital Image Processing*, New York: Addison-Wesley.

KILGER, M., 1992, Video-based traffic monitoring, in *IEEE 4th International Conference on Image Processing and Its Applications, Conference Publication no. 354*, Venue, Netherlands, pp. 563–6.

KING, D., 1991, Determination and reduction of cover type brightness variations with view angle in airborne multispectral video imagery, *Photogrammetric Engineering & Remote Sensing*, **57**(12), 1571–7.

LAWSON, D. R., GROBLICKI, P. J., SEDMAN, D.H., BISHOP, G. A. and GUENTHER, P. L., 1990, Emissions from in-use motor vehicles in Los Angeles: a pilot study of remote sensing and the inspection and maintenance program, *Journal of the Air Waste Management Association*, **40**(8), 1096–1105.

LILLESAND, T. and KEIFER, R., 1987, *Remote Sensing and Image Interpretation*, 2nd Edn, New York: John Wiley.

MAUSEL, P. W., EVERITT, J. H., ESCOBAR, D. E. and KING, D. J., 1992, Airborne videography: current status and future perspectives, *Photogrammetric Engineering & Remote Sensing*, **58**(8), 1189–95.

MCELHOE, H. B. and CONNER, W. D., 1986, Remote measurement of sulphur dioxide emissions using an ultraviolet light sensitive video system, *Journal of the Air Pollution Control Association*, **36**, 42–7.

MICHALOPOULOS, P. G. and WOLF, B., 1990, Machine–vision system for multispot vehicle detection, *Journal of Transportation Engineering*, **116**(3), 299–309.

ROURKE, A. and BELL, M. G. H., 1992, Wide area pedestrian monitoring using video image processing, in *IEEE 4th International Conference on Image Processing and Its Applications, Conference Publication no. 354*, Venue, Netherlands, pp. 563–6.

STEPHENS, R. D. and CADLE, S. H., 1991, Remote sensing measurements of carbon monoxide emissions from on-road vehicles, *Journal of Air Waste Management*, **41**, 39–46.

VLCEK, J., 1988, Nature of video images, in *Proceedings of the First Workshop on Videography, American Society of Photogrammetry and Remote Sensing*, Falls Church, VA.

WEIL, J. C., SYKES, R. I. and VENKATRAM, A., 1992, Evaluating air-quality models: review and outlook, *Journal of Applied Meteorology*, **31**, 1121–45.

WODTKE, M., 1993, *Mind over Media Creative Thinking Skills for Electronic Media*, New York: McGraw-Hill.

ZANNETTI, P., 1990, *Air Pollution Modeling, Theories, Computational Methods and Available Software*, New York: Van Nostrand Reinhold.

ZHANG, Y., STEDMAN, D. H., BISHOP, G. A., GUENTHER, P. L., BEATON, S. P. and PETERSON, J. E., 1993, On-road hydrocarbon remote sensing in the Denver area, *Environmental Science & Technology*, **27**, 1885–91.

CHAPTER SIX

CoastMAP: aerial photograph based mosaics in coastal zone management

Teresa Romão and António S. Câmara
Environmental Systems Analysis Group, New University of Lisbon, 2825 Monte de Caparica, Portugal

and

Mathilde Molendijk and Henk Scholten
Department of Regional Economics, Free University of Amsterdam, De Boelelaan 1105, 1081 HV Amsterdam, Netherlands

6.1 INTRODUCTION

The coastal zone is constantly changing, due to natural, social and economic processes. It contains some of the world's most productive and diverse resources, including extensive areas of mangroves, coral reefs and sea grass beds, which are highly sensitive to human intervention. These ecosystems are also the source of a significant proportion of global food production and support a variety of economic activities, including fisheries, tourism, recreation, industry and transportation. Due to such characteristics, which favor human settlement, about 60% of the world's population live in the coastal zone. Increasing pressures of rapid population growth, economic development and the resulting conflicting interests and competing demands for use of coastal areas and resources often call for a trade-off between conservation and development. Policy-makers are faced with the challenge of ensuring economic development while limiting the adverse impacts of such progress on natural areas. The coastal zone must be managed in such a way that the needs of today can be satisfied without endangering the opportunities of tomorrow.

Therefore, it becomes extremely important to develop coastal management systems that are able to analyze and control this evolution. Such a system should be simultaneously powerful and versatile. These characteristics are always difficult to combine, but we aim to achieve it by the integration of powerful tools with an intuitive and user-friendly interface.

Amongst the main objectives of the CoastMAP project, we can mention the following:

- exploration of the possibility of building databases behind multimedia applications
- exploration of the possibilities of integrating GIS functions into multimedia applications
- investigation of the role of exploratory spatial data analysis in the study of coastal zone phenomena

Based on the very important questions related to the future of our coastal zone and on the enormous potential offered by spatial information technology, a research program has been defined by the New University of Lisbon and the Free University of Amsterdam, in close collaboration with the Ministry of Traffic and Transport in the Netherlands (MD and RIKZ) and the Ministry of Environment in Portugal. The objective of this program is to develop a spatial information infrastructure which will improve the management of coastal zone, on the basis of the availability of strategic information which will make it possible to improve decision-making.

Several components of the spatial information infrastructure have been defined:

- the development of spatial databases of the coastal zone
- the development of videos to raise awareness of the possibilities of the role of spatial information technology for coastal zone development
- the development of multimedia GIS products to demonstrate the possibilities of such an infrastructure
- the development of user interfaces
- the development of spatial analysis tools
- the development of a World Wide Web server to exchange knowledge and information on the development of the coastal zone

This chapter describes a multimedia support information system for coastal zone management, named CoastMAP, which incorporates a wide variety of information from different databases. Our idea is to provide the user with data contained in distinct formats, giving him or her the possibility to combine this information. To allow the user to manipulate this data, GIS tools are available. Two methods are being developed to present the information: *flyover*, based on aerial photograph mosaic, and *hypermaps*.

Having access to such a large amount of information and powerful tools to consult, analyze and manipulate it, coastal zone managers can make more efficient decisions. Therefore, they can better explore coastal zone resources without endangering their future exploitation.

6.2 COASTAL MONITORING

The system currently being developed focuses on the coastal monitoring in the Netherlands.

Every year, coastal measurements are carried out along defined sections. These sections are delimited by beach posts located 200-250 m from each other. The results of these annual measurements are collected in the so-called JARKUS file. By analyzing this data, it is possible to keep track of coastline fluctuations. This method can be used to determine in which sections there is coastal accretion (sand sedimentation) and in which ones there is coastal recession (erosion). The main purpose of coastal policy is to preserve the 1990 coastline, the so-called basal coastline (Hillen, 1993). Annually, the coastline position is calculated on the basis of the results of the coastal measurements referenced above, related to previous years. The actual coastline position is then compared to the basal coastline to determine whether the latter has been crossed. If so, sand nourishment is executed. Moreover, it is possible to calculate the trend in coastline development and this way preview the year in which the basal coastline is likely to be crossed for a specific area (Van Heuvel, 1992).

The results of the annual measurements are recorded in the form of 'coastline charts', as shown in Figure 6.1. Each chart covers an area of 4 km × 4 km, so it takes 105 charts at a scale of 1:25 000 to cover the entire Dutch coastline (Van Heuvel, 1992).

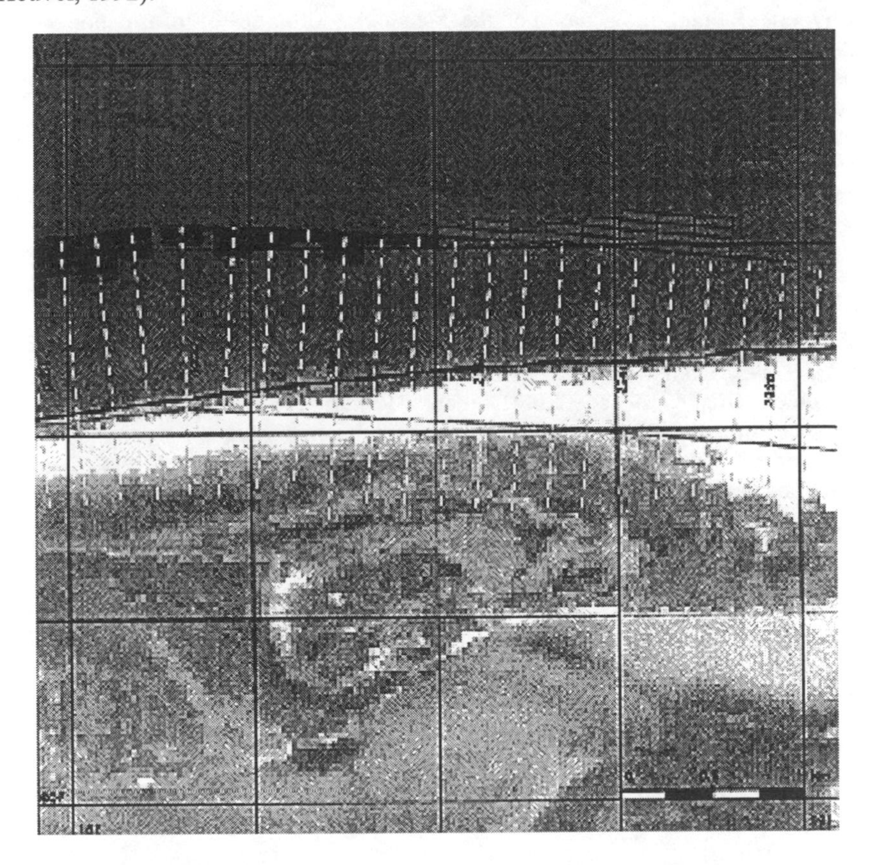

Figure 6.1 A 1992 coastline chart with a satellite photograph as the background.

This whole process is explained in the application that we are presenting in this chapter, using multimedia techniques. The data used in this system result from the assessment process described above, and were provided by the Dutch Ministry of Transport, Public Works and Water Management.

6.3 INFORMATION SYSTEM

CoastMAP is a multimedia GIS system based on aerial photograph mosaics, in which georeferenced information contained in internal and external databases is linked through a hypermedia structure.

6.3.1 Hypermedia

The hypertext technique is based on the interlinking of nodes of information which contain different media: therefore it is a natural environment for supporting multimedia interfaces (Nielsen, 1990). In this system the information can be presented through various layers (map, photograph and satellite image) that work as interactive images. Each layer is a node. The layers are connected to each other by hyperlinks (Figure 6.2). Therefore, the user can easily move between them. These layers can be seen separately or combined to reflect the relationship between data. Each layer is also linked to the database information. This hypermedia structure allows the user to access the same information via different paths.

An easy method of information retrieval is the use of hypermaps. A hypermap is a map that is hyperlinked with other information structures. It is a clickable map, from which the user can access all types of information, such as text, tables, images or even other maps (Raper, 1995). The different layers mentioned in the last paragraph are each hypermaps. Hypermaps are georeferenced. The data obtained when clicking on a hypermap is related to the clicked position.

Although there is no universally accepted definition, the multimedia concept is widely known as the integration of numerous, distinct media types simultaneously in one computer-based application (Lipton, 1992; Raper, 1995). The combination of the conventional textual information with images and sound can enrich the whole message to be transmitted. Multimedia allows the user to play an active role, controlling access to and manipulating an enormous amount of data (Ambron, 1988). Having access to photographs, animations and video, as well as to music, voices and meaningful sounds, the user has access to richer representations.

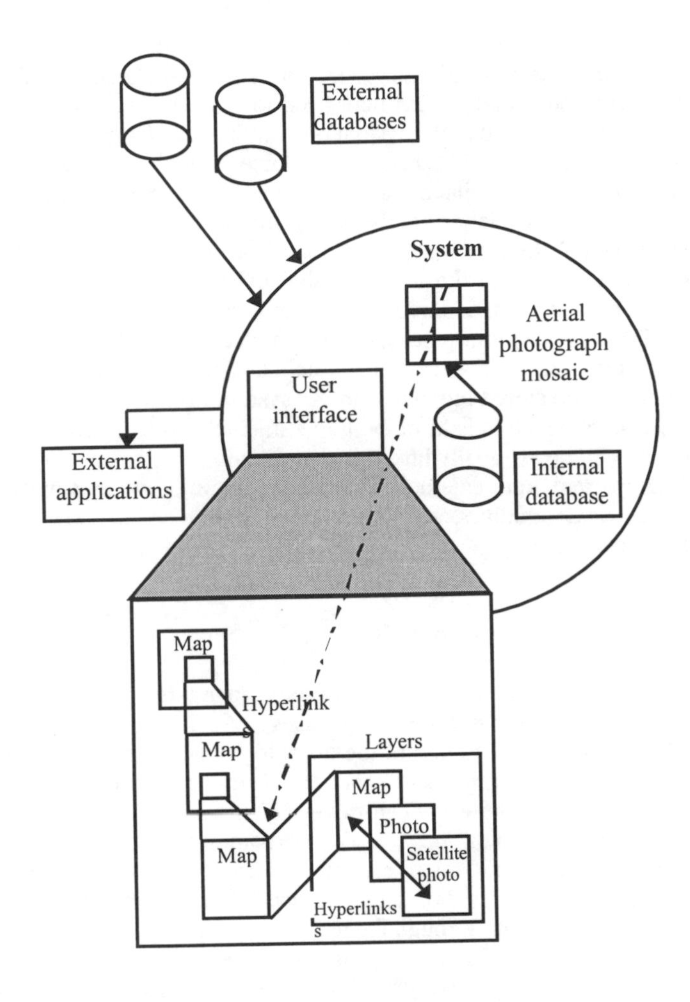

Figure 6.2 **The structure of the information system.**

Generated animation sequences will be also used to improve planning measures, allowing policy-makers to answer 'what if?' questions (Bill, 1994). In some specific situations, audio elements can be as important as visual elements, transmitting a sense of time, space, emotion or even reality (Burger, 1993). Meaningful sounds are used to immerse the user in the environment and voices are employed to reduce the time the user has to spend reading text. Both are even used as warning signs.

Sound is also integrated at key moments to reinforce and focus the user's attention on key scenes.

6.3.2 Linked databases

One of our main objectives is to explore the possibilities of having multiple linked databases behind the multimedia system. There is a large quantity of data that a user of such an environmental system may need to consult. These data can be stored in different databases linked to the system or even in a database built within the system. Our intention is to make all of this data available through one single interface.

Depending on the database structure involved in the process, it could be very difficult—even impossible—to get the data from each database directly. Our approach consists of having the data linked to the location to which it refers. In this way, the information is accessible through clickable hypermaps. A time attribute will also often be used for information retrieval, when data will be time- related. This enables a comparison between different scenarios which are separated by a time period. The data currently used in this system, which are related to the coastal monitoring measurements, are stored internally, in the system itself. This is only a subset of the whole database, which is held in the governmental services offices. Eventually, this information will be externally linked to the system.

The internal and external databases contain a wide variety of forms of information: text, graphics, satellite images, photographs, video, animations and statistical data.

6.3.3 Aerial photograph mosaics

The system presented in this chapter is based on aerial photograph mosaics which are linked to georeferenced multimedia databases.

The overall goal is to permit the users to visualize the whole coastline, and to have access, at the same time, to background information on what they are seeing. Sitting at their desks, the users can travel in virtual reality along the coast and see with their own eyes how it looks, with information reported to them about the area that they are flying over. This information is stored in multimedia linked databases, described above, which contain georeferenced data. These data are, therefore, connected to the photographs through the referencing coordinates.

This mechanism also includes a set of tools to navigate through the aerial photograph mosaic and to retrieve and manipulate the associated information. As in a plane, one can choose the flight direction, select a place to land and even fly at different altitudes, with differently scaled overviews of the zone being flown over. A zoom tool will be also provided.

It is also intended to provide a comparison between the actual images seen during the 'flyover' and old photographs of the area concerned, while simultaneously having access to numeric information and charts.

The coordinates will be permanently displayed on the screen and the user will have access to a map on which his or her position is highlighted (Figure 6.4). This map will also be clickable, thus enabling the user to choose another area to fly over. In this way, the user will never be able to get lost.

The development of the flyover mechanism implies the resolution of some previous problems, such as digitization and orthorectification of photographs. Scanning of photographs is only possible after deciding on the intended resolution.

This decision depends on the trade-off between the image's quality and size, according to the desired level of detail. Orthorectification is needed due to the fact that the Earth's surfacc is not flat. For this process, elevation information is essential. Sometimes it is also necessary to accomplish color correction. Photographs should have colors of similar tone, so that when they are placed together one cannot recognize the joining line. When ready, the aerial photographs should be stored in a database.

In the present state, the 'flyover' mechanism is being developed using 'coastline charts'. These charts constitute an in-line mosaic of satellite images or scanned topographic charts (in this case). Thus, it is perfectly suitable to play the role of an aerial photograph mosaic, since we are not concerned with the technical problems described above, but only with its end results.

6.4 SPATIAL ANALYSIS

The applications of GIS in coastal zone management lie mainly in the following areas (Martin, 1993):

- *Stock-taking and integration of various data.* GIS technology is very useful for quickly finding and sorting data that are based on the location and other spatial relations.
- *Analysis of coastal change.* With the help of GIS, changes in coastal systems can be monitored. For example, processes of coastal erosion and sedimentation can be analyzed, and sand replenishment programs can be evaluated. With the use of simulation processes, developments in the coastal area can be extrapolated from the past, as well as into the future (e.g. the rising sea level resulting from climatic change can be visualized). Simulations can also be used to illustrate the possible effects of various policy considerations.
- *Analysis of exploratory spatial data.* Spatial data are data that are associated with a location. Meaningful analysis of such data must be carried out with the help of maps. Exploratory data analysis is an interactive process of map browsing using statistical views in addition to direct examination of maps.
- *Coastal resource survey and management.* GIS are often used for exploration and exploitation (allocation) of natural resources such as oil and sea sand. GIS are also applied in the search for suitable locations. (For example: Which vulnerable dunes should be protected? What locations would be suitable for a fishery?)

Research has been carried out recently with the aim of linking multimedia features to GIS systems; namely, two projects carried out by the authors:

- On the basis of the GIS program Mapinfo, Geodan have developed an application that combines GIS techniques with multimedia data-like images and sound, providing the user with an interactive exploration of this data (Scholten, 1995).
- At the New University of Lisbon, the Environmental Systems Analysis Group, in cooperation with the Portuguese National Center for Geographical Information, are developing some GIS applications in which they introduce multimedia techniques in order to better explore diverse phenomena (Câmara, 1995).

A different approach is pursued here, in an attempt to introduce some GIS tools within a multimedia application.

Some GIS analyses, such as classification, overlay or neighborhood, will be available. These will enable the user to manipulate the information in order to obtain the needed data. These GIS tools will be manipulated through a toolbox that is described later in this chapter. The user will be able to make management choices, and instantly view the results and impacts of these choices.

6.5 USER INTERFACE

An important issue is the design of the interface, which must be as intuitive and user-friendly as possible. In this way, the information will be organized in a hypertext structure and will be easily accessible through menu options, interactive images and a toolbox that will enable the user to select the different tools available.

The hypermap's mechanism is used to drive the user into more detailed maps: from a country area to a beach area, and through its province area (Figures 6.2 and 6.3) (Câmara, 1991). At each level of detail, the user can have access to the corresponding detailed data. To achieve that, the user can choose between two different work modes: *zoom* and *spy*. The *zoom* mode allows the user to travel into more detailed maps which lead to more detailed data. The *spy* mode enables the user to consult data related to the current level of information. For example, when the user is confronted with a map of the Netherlands and chooses the *spy* mode, he or she can access information on the number of Dutch people living in the coastal zone. Otherwise, if the user wishes to have detailed information about a specific area, he or she should select the *zoom* mode and click on that area to obtain a map of the area, and then have access to the related detailed data (Figures 6.2 and 6.3). The *zoom* mode is bi-directional: there is a *zoom in* and a *zoom out*, in order to go into a detailed view and to come out to the more general overview.

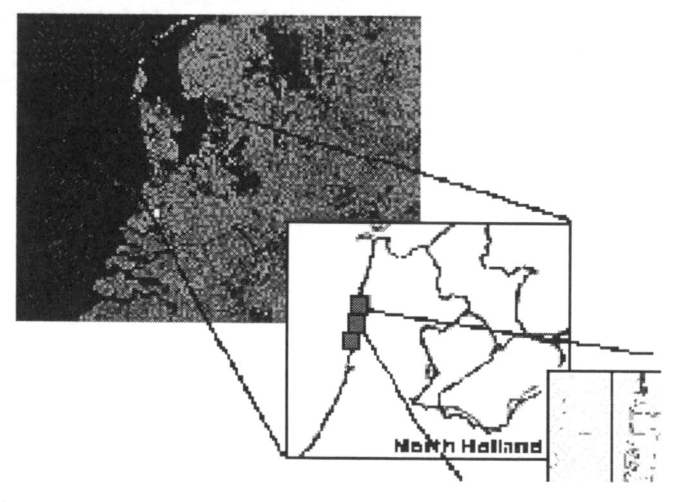

Figure 6.3 **Hypermaps. The user can move between different levels of detail.**

When exploring detailed data, such as the 'coastline charts', the user can make use of a set of available tools from a displayed toolbox (Figure 6.4). Besides the DOS shell and print screen options, these tools will include data manipulation devices, such as GIS functionalities, calculation, selection and network tools. It is also possible to execute calls to external applications which allow the user to visualize and manipulate data that is in some specific format. Each tool is represented by an icon that should clearly show that tool's functionality.

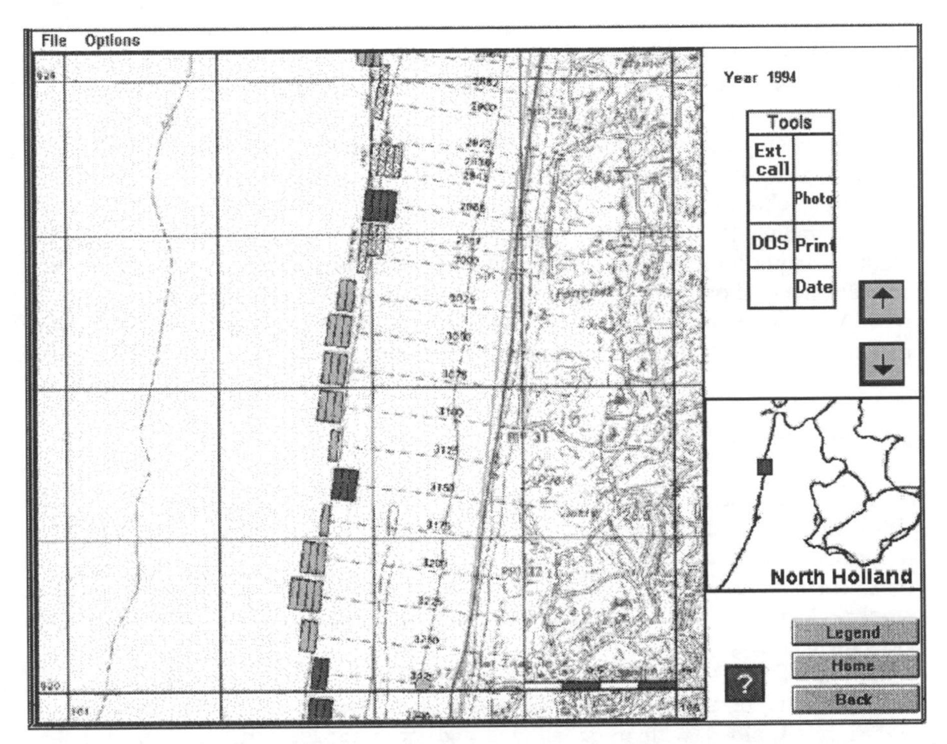

Figure 6.4 The CoastMAP interface. The user's position is highlighted on a small-scale map. Some tools are available from a toolbox. The user can 'fly over ' the 'coastline charts' and retrieve information when clicking on a specific point.

On-line help will be always available, as well as context-sensitive help for every feature of the system (Sellen, 1990). By clicking on a button, the user can have access to a dynamic overview showing the system structure, on which his or her current location will be indicated, as well as the most recently visited screens. Due to the high number of nodes, the diagram will only be displayed in detail for the area surrounding the user's current position (Nielsen, 1990). The user can always return to the introductory screen, where a schema representing the whole process of coastal monitoring is shown. Here, one can select one of its steps. Custom buttons, such as help, back or home buttons, are always available at the same position on the screen, so that users will quickly learn their positions and always know where to find them.

Sound warnings are used to call the user's attention to special points during the data analysis and exploration. Voice annotations substitute for large amounts of text that will not be read by the user. Video sequences are available to complement text or vocal explanations.

There is also a language option that allows the user to select the language that he or she prefers to use when exploring the application. The available languages are Dutch, English and Portuguese. Since icons can intuitively express their meaning, being independent of language, icon design and usability are also being studied. To be useful, an icon should be graphically clear and semantically unambiguous. Some features can be explored when designing icons, such as colors, movement or background (Yazdani, 1993). Colors can be used as coding, although it is not possible to attribute a single meaning to a color: for example, red is often associated with danger, but it is also frequently used to attract the user's attention in situations not related to danger (Salomom, 1990). Movement is sometimes helpful to explain the significance of an icon, or to signal that there is an animation behind it.

Morphing techniques are also used to help coast managers to visualize the coast's morphological changes. Based on old photographs or maps, morphing movies clearly show the evolution in the coastal zones for a specific area.

A prototype of the application will be tested by coastal zone managers, with special emphasis on the design of the interface and the selection of available tools.

6.6 FUTURE WORK

Future developments will be related to the expansion of the existing system, the publishing of the hypermedia document, and its future extension for computer-supported collaborative management.

The expansion of the current system will focus on the addition of new data types and analytic functions. Thus the use of stereoscopic images and 360 degree views will be considered.

The CD-ROM is a unique publishing medium, designed to deliver all types of digital data: text, graphics, sound, video, photographs or software. Its durability and stability, along with its small size and large capacity, make it an excellent instrument with which to deliver a large amount of digital data (Nadeau, 1994). In addition, a large number of CD-ROM users no longer limit the portability of this easy-to-use medium. For all of these reasons, we plan to create a CD-ROM to distribute the final system. It seems to be the most suitable vehicle, considering the type of data used in the system and especially the aerial photographs.

Furthermore, we will investigate the evolution of the project into a computer-supported collaborative management system. An environmental decision commonly depends on the knowledge, experience and investigation of many people with differing backgrounds. In the coastal monitoring process, for example, decisions are taken by provincial consultative bodies, which include people from the state departments, water boards, provincial authorities and sometimes from coastal municipalities and nature conservation organizations. In this way, it would be very helpful if all people involved could work together and discuss their ideas without having to leave their offices.

6.7 CONCLUSIONS

As a multimedia decision support system for coastal zone management that includes spatial analysis tools, CoastMAP aims to facilitate the study, analysis and exploration of coastal areas.

The multimedia information system relies upon mosaics of aerial photographs linked to georeferenced multimedia databases. These databases include video images on the dynamics of coastal processes.

Using CoastMAP, coastal zone managers will have access to aerial photograph mosaics and may fly over the coast, being supplied with information related to the area that they are flying over. In this way, users will feel more deeply involved in the natural environment they are analyzing, which facilitates the perception of the problem.

Future developments will include the system's expansion at both the data and analysis levels, publishing in CD-ROM format and the development of computer-supported collaborative management tools.

ACKNOWLEDGMENTS

We thank Tjark van Heuvel for the data provided and for feedback. We also thank the reviewers for their work and their excellent feedback.

The work of Teresa Romão is funded by PRAXIS XXI through a PhD fellowship associated with the project PRAXIS/3/3.2/AMB/04/94.

REFERENCES

AMBROM, S. and HOOPER, K., 1988, Interactive multimedia: visions of multimedia for developers, educators & information providers, washington, dc: microsoft press.

BILL, R., 1994, Multimedia GIS, definition, requirements and applications, *Gisdata Reading*, Rostock.

BURGER, J., 1993, *The Desktop Multimedia Bible*, Reading, MA: Addison-Wesley.

CÂMARA, A., 1995, Tutorial of multimedia and GIS (unpublished notes), *Joint European Conference of Geographic Information*, The Hague.

CÂMARA, A., GOMES, A. L., FONSECA, A. and LUCENA E VALE, M. J., 1991, Hypersnige, a navigation system for geographic information, in *Proceedings of the European Geographic Information Systems Conference*, Brussels, Belgium.

HILLEN, R. and DE HAAN, TJ., 1993, Development and implementation of the coastal defence policy for the Netherlands, in HILLEN, R. and VERHAHEN,

H.J. (Eds.), *Coastlines of the Southern North Sea*, New York: ASCE.

LIPTON, R., 1992, *Multimedia Toolkit*, New York: Random House.

MARTIN, K. ST., 1993, Applications in coastal zone research & management, in *Explorations in Geographic Information Systems Technology*, UNITEAR, Vol. III.

NADEAU, M., 1994, *The Byte Guide to CD-ROM*, Berkeley, CA: McGraw-Hill.

NIELSEN, J., 1990, *Hypertext & Hypermedia*, San Diego: Academic Press.

RAPER, J., 1995, Prospects for spatial multimedia, unpublished manuscript, London: Birbeck College.

SALOMOM, G., 1990, *New Uses for Colors*, in LAUREL, B. (Ed.), *The Art of Human-Computer Interface Design,* Reading, MA: Addison-Wesley.

SCHOLTEN, H., 1995, Tutorial of multimedia and GIS (unpublished notes), *Joint European Conference of Geographic Information,* The Hague.

SELLEN, A. and NICOL, A., 1990, Building user-centered on-line help, in LAUREL, B. (Ed.), *The Art of Human-Computer Interface Design*, Reading, MA: Addison-Wesley.

VAN HEUVEL, TJ. and HILLEN, R., 1992, *Coastline Management*, Rijkswaterstaat, Tidal Waters Division, The Hague.

YAZDANI, M., 1993, *Multilingual Multimedia: Bridging the Language Barrier with Intelligent Systems*, Oxford: Intellect Books.

CHAPTER SEVEN

The EXPO'98 CD-ROM: a multimedia system for environmental exploration

A. Fonseca, C. Gouveia and J. P. Fernandes
Centro Nacional de Informação Geográfica,
Rua Braamcamp, 82, 5° Esq., 1200 Lisboa, Portugal

and

A. S. Câmara, A. Pinheiro, D. Aragão, J. P. Silva and M. I. Sousa
Grupo de Análise de Sistemas Ambientais, Faculdade de Ciências e Tecnologia,
Universidade Nova de Lisboa,
Quinta da Torre, 2725 Monte de Caparica, Portugal

7.1 INTRODUCTION

Environmental education and awareness are major concerns in the development of environmental policies. They require the existence of means that allow a better understanding of environmental phenomena and about the key issues involved in the definition of strategies and the implementation of environmental management measures.

Our goal is to design an environmental exploratory system using interactive technologies, to address the above-mentioned problem. EXPO'98 was an excellent opportunity to test these ideas, because the EXPO'98 site is an area in which several environmental problems needed to be solved. The environmental studies on the site have produced a considerable amount of information, including environmental, geographic and audio-visual data, and represented a valuable source of data for the development of such an application.

The environmental exploratory multimedia system for EXPO'98 in Lisbon presented environmental information about this area, during the different implementation phases of the project, thus enabling visitors to explore the spatial nature of environmental problems. The system is available on CD-ROM, and an HTML version is also available on the Internet.

The development of this multimedia system that allows access to environmental information about EXPO'98 in Lisbon aims to improve access to the environmental information about that area, to provide new ways of explaining environmental phenomena and, overall, to increase public awareness and public participation in the environmental decision-making process. It has both educational and entertainment aspects, exploring the interactive capabilities of multimedia technologies.

Previous work includes the use of interactive graphical systems for environmental applications, such as, for example: the work developed by Loucks *et al.* (1985) on the use of interactive water resources modeling systems; the research of Arnold and Orlob (1989) on the development of decision support systems for estuarine water-quality management; the work of Fedra (1993) on the development of interactive environmental software; and the application Hypertejo, an hypermedia system to explore watershed information, developed by Câmara (1989).

Several multimedia applications have been developed in different fields, as reviewed in Raper (1997), Khoshafian and Baker (1996), Gibbs and Tsichritzis (1995) and Lang (1992). Some of these developments can be mentioned in this context.

In the end of the 1970s, a pioneer project took place at MIT, the Aspen project, in which film shots taken from a moving vehicle traveling through the town of Aspen were stored on videodisks and then accessed interactively to simulate driving through the town.

There was also a classic pioneering example of integration of multiple data types in the form of the BBC Domesday project (Rhind *et al.*, 1988). The Domesday system can be considered to be the earliest hypermedia spatial database: it contained a huge amount of spatial data relating to population, employment and the environment, with Ordnance Survey maps and photographs. Other works can be also mentioned as multimedia spatial simulation systems, such as SimCity and SimEarth, for urban and global modeling (Joffe and Wright, 1989), and the Picture Simulator (Câmara *et al.*, 1994), for environmental data sets.

More recent developments, in multimedia applications to urban planning, include the use of sound and video to improve the analysis of traffic problems and noise assessment. In these developments, speech recognition and voice annotation are used to facilitate public participation during the planning process.

One of the most significant developments of this kind is the Collaborative Planning System (CPS; Shiffer, 1993), based on SuperCard and using Quicktime, designed to evaluate planning scenarios for Washington, DC, and allowing users to navigate through a wide range of multimedia data.

The explosive growth of the Internet has pushed multimedia applications from stand-alone versions toward network-based systems (Nielsen, 1995). World Wide Web (WWW) multimedia systems are emerging and will probably, in the near future, become one of the most important supporting technologies for environmental and planning applications (Shiffer, 1995b). Networked systems facilitate information gathering and updating, improve communication and cooperation among professionals, and allow increased public participation in the contexts of environmental and planning decision-making.

Several multimedia applications, some of which are already exploring the advantages of the WWW, are being developed in the environmental and planning fields. These applications address different issues, and include, among others, data collection tools (Ferreira, 1995), multimedia kiosks and CD-ROMs for retrieval of information (Blat, 1995; Fonseca *et al.*, 1995), spatial data infrastructures (Henriques, 1996; Geodan, 1997), image-based multimedia systems (Romão *et al.*, 1995; Smith and Frew, 1995; Fernandes *et al.*, 1997), collaborative planning systems (Carver, 1995; Shiffer, 1995a; Gordon and Karacapilidis, 1996) and public participation systems (Gouveia, 1996; Ferraz de Abreu, 1997).

The following section presents the requisites for an environmental exploratory system based on analysis of the nature of environmental problems and its information and manipulation requirements. The application to EXPO'98 is then discussed and illustrated, in terms of interface design, available data and interactive manipulation capabilities. Finally, future developments in the use of multimedia technologies for this kind of application are discussed.

7.2 REQUISITES FOR AN ENVIRONMENTAL EXPLORATORY SYSTEM

The use of multimedia technologies for environmental exploration is closely associated with the nature of environmental problems. The main features of environmental problems are discussed in the following sections, after which an analysis of hypermedia systems and their role in cognition processes is presented. According to those features, a presentation of the major requisites for a multimedia environmental exploratory system is then made.

7.2.1 The nature of environmental problems

In essence, environmental problems can be described as multidimensional, involving several components, variables and space-time scales. On the other hand, the perception of these problems has strong visual and audible features, and a highly dynamic nature.

This dynamic aspect is possibly the major feature of environmental problems; hence the importance of being able to portray it. Simulation models have an important role to play in dealing with this dynamic nature.

Another essential aspect of environmental problems is their spatial dimension. Environmental problems deal with spatially varying phenomena with no, or very unclear, borders. For example, circulation in the atmosphere is a typically unconfined three-dimensional problem, and the circulation of a pollutant in a reservoir takes place in three dimensions (Guariso and Werthner, 1989). Modeling capabilities try to deal with this spatial nature, and the use of spatial information systems such as geographic information systems (GIS) also helps to incorporate this characteristic in environmental analysis.

Another characteristic of environmental problems is complexity in terms of the multiplicity of components, variables, criteria or points of view under which they can be seen and of the involvement of different social groups. Additionally, there is a need for a great deal of data in order to properly model and verify environmental problems. There is a massive data requirement, due to the complexity of the problems already mentioned, and a need for careful experimentation before a model is actually deployed.

Most environmental problems have a strong visual and audible nature. For example, analysis of the air pollution due to plume propagation can benefit from access to video images at regular time intervals. Noise assessment can be improved considerably if sound representations are associated with the other kinds of data.

The analysis of environmental problems implies consideration of the interrelationships between phenomena and compartments. To facilitate this global analysis, the possibility of navigating interactively through the real environmental universe and its symbolic representations can be highly significant.

In the visualization of natural phenomena, the use of tools such as GIS, incorporating realistic images or their representations (e.g. maps and graphics), normally requires a great deal of expertise in order to take advantage of their full capabilities. More recently, the use of multimedia applications shows that the general public is becoming quite used to hypertext concepts. For example, the use of multimedia kiosks, games, CD-ROM titles and the World Wide Web (WWW) is now widespread.

7.2.2 Hypermedia systems

Hypermedia systems, being based on a structure of nodes and links, may contribute to the development of applications where the user can freely and intuitively explore a set of data.

The design of hypermedia systems is driven by technological innovations and user-oriented issues, associated with cognition and human information processing (Thuring *et al*., 1995). Two kinds of applications can be identified in association with cognitive aspects within hypermedia systems (Stotts and Furuta, 1991):

- The explorer approach, where the user is encouraged to wander through large sets of information, gathering knowledge along the way. These applications can be seen as browsable databases that can be freely explored by the user.
- The document-centered approach, which is more directly tied to specific problem-solving, being much more structured and constrained. Examples of these applications can be translated in hyperdocuments that guide the user through the information with a specific intention, along a pre-defined structure.

The first approach fits better with unconstrained search and information retrieval, and the second one is more suited for tasks that require deep understanding and learning.

In both situations, the aim of the hypermedia applications is to contribute to improving comprehension. Coherence is a crucial factor in this context. The user's ability to understand and remember a subject depends on its degree of coherence at the local and global level. Different cues can be used to guarantee these different levels of coherence. For example, at the local level the fragmentation characteristic of hypertext should be limited, because it can result in a lack of interpretative context. At the global level, cues should help the user to identify the major components of the hypermedia application and the way in which these constitute its overall structure.

In order to reduce the 'effort and concentration necessary to maintain several tasks or trails at one time' (Conklin, 1987) which is associated with the limited capacity of human information processing, some other cues should be considered, which are associated with orientation, navigation and user-interface adjustment.

Orientation and navigation concepts are related to the 'travel metaphor', which is frequently used in these systems. Orientation cues include the ways to: identify the current position with respect to the overall structure; reconstruct the route that led to that position; and distinguish among different options for moving on from this position (Thuring *et al*., 1995).

While orientation facilities help users to find their way, navigation facilities enable users to make their way. The navigation cues are based on two crucial aspects: direction and distance. Through direction, the user is able to distinguish forward and backward navigation, either on the horizontal or the vertical dimension. Through distance, it is possible to consider steps and jumps. Steps correspond to following nodes that are directly linked. A jump refers to the performance of a link to a node that is not immediately connected to the current position.

User-interface adjustments are intended to eliminate the dispensable activities that can negatively influence user performance (for example, the necessity to manually close windows on the screen).

The consideration of cognitive design principles is thus a determining factor in obtaining successful multimedia applications.

7.2.3 Environmental exploratory systems

Multimedia capabilities can be explored within environmental exploratory systems to facilitate access to environmental information, to improve the explanation of environmental phenomena and to heighten the perception of environmental processes.

However, to better understand the design of such a multimedia environmental exploratory system, one should analyze its requisites according to the previously presented nature of environmental problems. Thus the system should be able to deal with the dynamic nature of environmental problems, their spatial dimension, the high level of complexity and interrelationships that take place, the strong visual and audible nature of such phenomena and problems, and the large data requirements.

For the analysis of requisites, three main issues are considered in the development of such a system: the data, the models and the user interface.

The data

Such a system comprises different types of data: maps and aerial images at different scales and from different dates; 3-D models that represent existing or future infrastructures; environmental data concerning, for example, field measurements of pollutants; and visualizations of these data, surface photographs, videos and sounds, as well as realistic views of specific sites.

The incorporation and manipulation of videos, images and sounds with spatial data handling functionalities may facilitate the consideration of environmental time and space dimensions. Video and sound are dynamic components, and their manipulation in association with spatial data can contribute to improving environmental analysis. For example, access to videos associated with specific directions or around specific points (360 degree pan views) on a map, at different times, gives a more realistic and intuitive view of a certain development or phenomenon. To fly over an area using the aerial photographs, topographic information and animation capabilities of these systems can give a much more realistic and dynamic view of a study area.

The integration of multiple types of data and exploration of the interactive capabilities of multimedia technologies allows users access to multiple views of the same reality that can be interactively explored. This can contribute to better handling of environmental complexity, and can facilitate the understanding of interrelationships. This is very important as far as the communication of environmental information to the public is concerned, because more realistic and dynamic views of the data, and of the results of environmental impact analyses, can be achieved. For example, impacts on water quality from a certain development might be much more easily transmitted to the public via images or videos that show the changes in water use, rather than through the presentation of numeric tables or graphs of water-quality data and standards.

The manipulation of images, videos and sounds can obviously help in tackling the visual and audible nature of environmental phenomena. Morphing operations can visually transmit the evolution of some phenomena, allowing a better understanding of spatial and temporal trends. Visual simulation models can be used to assess different land occupation alternatives. Sound can be used to represent numeric data, as, for example, in the case of the spatial distribution of noise.

The use of these additional types of data can contribute to increasing the data available for the analysis of environmental problems. For certain applications, video and images can be as important as a map or access to a database.

The models

Simulation models play an important role in dealing with the dynamic nature of environmental problems. These models should be able to tackle the spatial and temporal dimensions of these phenomena.

Assessment of the impacts of human activities makes the prediction and analysis of environmental impacts and risks the basis for rational management. Thus an environmental exploratory system should include the available numeric environmental simulation models for water, air, noise, soil, biota, landscape and infrastructure.

On the other hand, pictorial simulation models should be also considered, as they can take advantage of the visual components of these systems, thus allowing a more natural and intuitive way of assessing environmental phenomena.

The user interface

The design of the interface for such a system should take into consideration several topics, such as: the functional requirements of the system according to the tasks to be performed; the model's adaptation to the user's cognitive representations; and the definition of the types of dialog with the user.

Perceptual problems, which are important in a multimedia environment, are related to the resolution, color and sound, and to the response time. Some of these topics have been considered in the development of video games (Crawford, 1990), a field in which user interface issues have been carefully taken into account.

Interactivity is another determinant characteristic of the user interface. Multimedia models have to be developed to support a high degree of interactivity, which may be measured by its frequency, range and significance (Laurel, 1990).

Following these principles, this kind of user interface should give the user an improved way of navigating through the environmental data and interacting with it. It should explore the use of specific cues to guarantee consistency, orientation and navigation within the system, in an intuitive way (Shneiderman, 1998).

7.3 EXPO'98 ENVIRONMENTAL EXPLORATORY SYSTEM

From May to September 1998, the last world exhibition of the century, entitled 'The Oceans—A Heritage for the Future', took place at a riverside site in eastern Lisbon.

Within this area, there were, for a long period of time, several industrial facilities, such as oil refineries, gas deposits, a sand extraction plant, a landfill site and a wastewater treatment plant. Their activities had severe negative environmental impacts, such as contamination of the soil and of sediments, water pollution, the release of odors and the proliferation of rodents. For this reason, this part of the city had become one of the most degraded and spatially segregated areas of Lisbon.

The thorough urban renewal program that was submitted for this site required major changes in the existing land use and the resolution of various environmental problems. The environmental remediation process involved a massive data collection and production exercise, which was essential in the assessment of the EXPO'98 environmental impact.

A CD-ROM containing these data about the environmental remediation process was produced. It enables exploration of the environmental story of the area, thus providing better knowledge of and more information on environmental issues.

The initial developments within this application were primarily concerned with the data contents, structure and user interface design aspects. The data contents are related to the selection of the information that should be included in the application; data structure and user interface design deals with the method of organizing the application contents, and with how to present structural and content information to the users. The next sections will describe each of these development efforts and its main results in terms of the final product.

7.3.1 Data contents

In the EXPO'98 Environmental Exploratory System, various types of sources were used for environmental data acquisition.

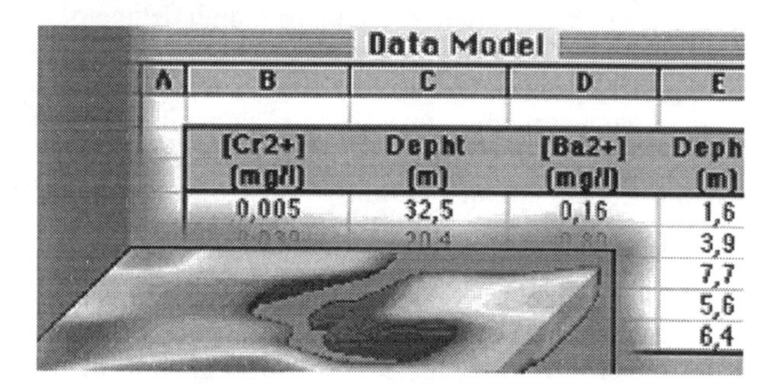

Figure 7.1 Data visualization operations.

Numeric data, collected in field studies and monitoring activities, were stored in a relational database and used to produce new kinds of multimedia information (e.g. maps and 3-D visualizations), as presented in Figure 7.1.

Textual data, collected from environmental baseline studies and sectorial reports, were embedded with hypertext functionalities (e.g. links to an environmental glossary or links to contextually different media). Visual (still photographs, aerial imagery, maps, animations and video clips) and acoustical (sound tracks and narration) data were stored using the most convenient compression method and indexed for later retrieval. Compression methods were of major importance due to the limitations of physical space on CD-ROM disks.

7.3.2 Structure and design of user interface

The general metaphor used for this environmental exploratory system aimed at providing an interactive and exploratory journey through the EXPO'98 site. A transparent air bubble was created, from which the user could direct his or her actions (Figure 7.2). As in a true vehicle, the air bubble has a control panel, the TABLIER, which is always partially visible on the screen. The screen constitutes the background scene, and the TABLIER is the interface device for interaction (Figure 7.3).

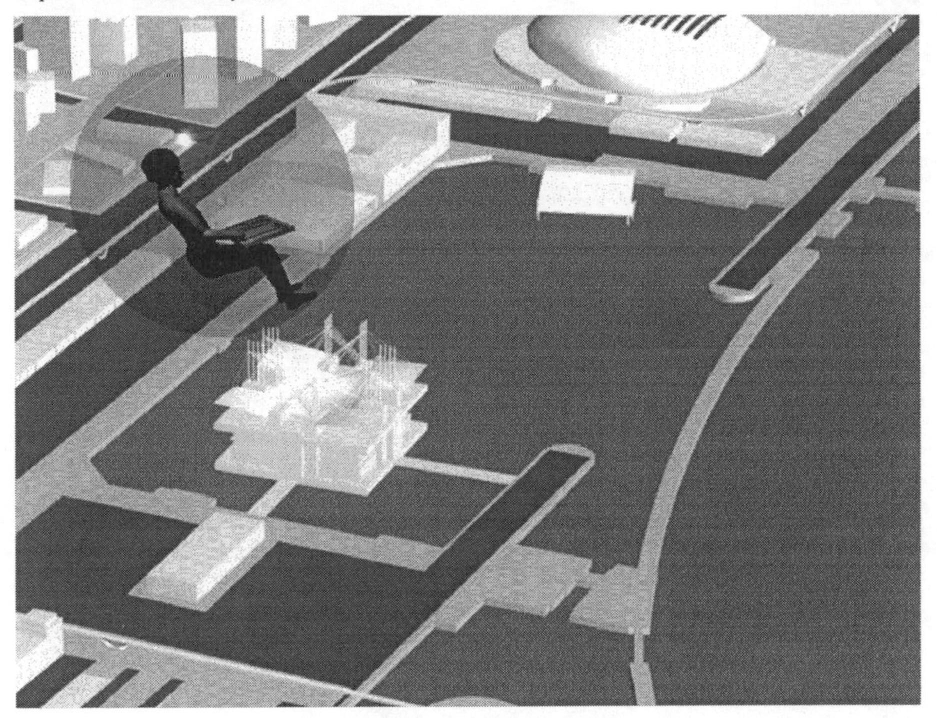

Figure 7.2 Navigation through the system based on the transparent vehicle.

The TABLIER has several environmental exploration features that will allow the user to perform his or her exploratory journey, including navigation and orientation facilities, visualization and simulation tools, and an environmental glossary.

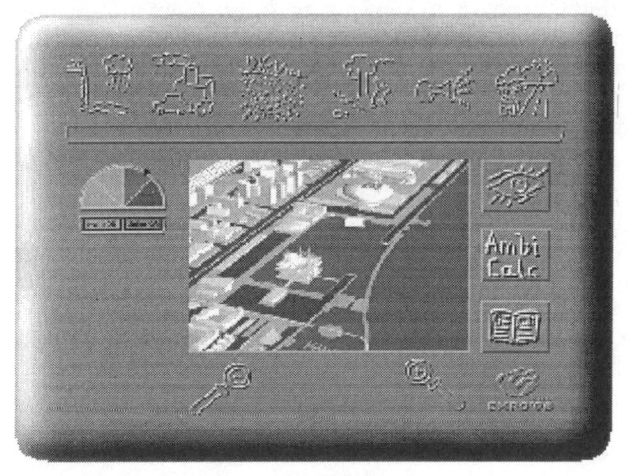

Figure 7.3 The TABLIER.

Navigation facilities

The navigation module is made up of several complementary navigation facilities that will help users to make their way through the exploration process. Navigation in the EXPO'98 Environmental Exploratory System takes into consideration the spatial, temporal and thematic dimensions. Each of these dimensions has its own independent structure, according to the available historical environmental data contents. The navigation process in any of these dimensions involves the definition of a direction and a distance path.

With respect to direction, the user is able to perform forward or backward navigation. The former is the case when a user is moving to a place where he or she has not been before, seeking for new information; while the latter is the case when a user is moving to a node that has previously been browsed, trying to retrieve information that is already known.

Regarding the distance definition, the user will be able to perform navigation steps or navigation jumps. Steps are simple moves from one place to another in its direct neighborhood. Jumps occur whenever a user wants to reach a place that is not directly connected to its current position. Jumps are of major relevance in the case of backward navigation to reach some formerly accessed information, after having traveled a considerable distance.

Spatial navigation is performed both by flying over maps or aerial imagery and by walking through 3-D models of specific places. Because environmental data (e.g. ground photographs, video clips and numeric data) have a spatial component, for each spatial step that a user performs the information contents are automatically changed. Zooming facilities, to distinguish between local, regional and global variations, are available because of the step and jump navigation capabilities.

A time slider, accessible through the TABLIER, was specially designed in order to perform temporal navigation. Four main temporal periods were considered: baseline conditions, the construction phase, the exhibition phase and the post-exhibition phase. In some cases, where more information is available (due to the monitoring program for some environmental components), it will be possible to perform smaller steps. The ability to perform temporal steps and temporal jumps gives the user the opportunity to analyze the dynamic evolution of environmental data sets over time.

Finally, thematic selections are possible, so that the system retrieves only the information related to the specific environmental component selected. The thematic structure considers not only (1) the physical and chemical characteristics of the environmental components (e.g. surface water quality, or the atmospheric concentration of different pollutants) but also (2) biological conditions (e.g. the diversity of fauna and flora) and (3) the cultural factors (e.g. land use, aesthetics, recreation, man-made facilities and activities) involved in the EXPO'98 Environmental Remediation Process.

Thematic navigation will follow a traditional node-link structure, providing the user with the ability to access the desired environmental feature for spatial and temporal exploration. This document structured-oriented scheme is necessary to guarantee users acquisition of the thematic contents and their interactions.

This scheme enables the user to explore the spatial and temporal patterns of each environmental information unit, along the thematic navigation process. The thematic interactions (e.g. pollution loads and water quality parameter values) can be readily apprehended because of the links that tie them together.

The navigation scheme described above has a free exploration nature, which contrasts with document-oriented hypermedia structured systems, in which a rigid sequence and a number of constraints are used to guide users in their navigational process. In fact, the ability to navigate through various time, space and thematic dimensions gives the user an unique exploratory freedom. The combination of these three different dimensions provides access to different views of the same scenario, that in some way may contribute to an overall understanding of environmental phenomena.

Orientation facilities

In order to deal with the sense of disorientation that these systems may cause, some orientation features were added. These orientation features deal mostly with the incorporation of a color scheme in the structure representation, in order to provide the user with a permanent orientation in the three dimensions considered.

Although maps, aerial imagery and 3-D models, as spatial navigation facilities, provide orientation and context to the other navigation activities, spatial structure representation is embedded in some orientation facilities to help the users find their way through the navigational process. These orientation facilities include the following:

- *Spatial navigation status*: a map in the TABLIER that will represent the spatial locations that have not yet been explored. These spatial locations are represented by blue areas. Red areas indicate the actual spatial position of the user. Purple areas represent the path followed by the user in his or her spatial navigation.
- *The zooming facility*: this feature, that allows the user to navigate at different scales (local, regional or global level) has an influence on the type of spatial structure representation.

Temporal structure representation is intrinsically incorporated in the time slider design. Located in the TABLIER, this time slider always indicates the actual status of temporal navigation. The same color scheme is applied; but now, instead of mapping representations, the user can find his or her way by labeling references in the time steps considered.

The thematic structure representation follows the same approach. The selected theme and sub-themes (e.g. surface water quality, water-quality parameters or water-quality standards) associated with a certain time-space combination are always identified in the TABLIER. Colors are also used to allow the identification of themes that have already been browsed for that space and time setting, and those that are still available for exploration.

Since both information contents (time, space and theme) and structure references (orientation facilities) are simultaneously present, users are constantly supported in their cognitive process of information acquisition and comprehension.

It is, therefore, possible to select to fly over an aerial photograph of 40 years ago and compare it with the one from 1994, and then land on the ground and walk through a 3-D model of that site, selecting and retrieving information from the existing 3-D objects (natural or man-made) for different environmental components. One can also dive into the polluted waters of the Trancão River, and find where the main sources of pollution are located, guided by the water-quality indicator on the TABLIER.

The visualization tool

A VISUALISER is available in the TABLIER, which allows access to 3-D visualizations, morphing operations, animations, graphics, photographs and videos (Figure 7.4). The VISUALISER also gives access to tables with environmental data and standards from the EXPO'98 Environmental Database, which are the basis of the visualization operations.

For example, while exploring the EXPO'98 site in the transparent air bubble, the user may find the need to gain an overview of the environmental quality. In the TABLIER, there is an area specially dedicated to present environmental indicators, one for each environmental component, giving a brief idea of the quality of that component for that given time and space setting.

Figure 7.4 The VISUALISER—a visualization tool that gives access to 3-D visualizations, morphing operations, animations, graphics, photographs and videos.

More detailed information on specific environmental components is available, including 3-D visualizations of environmental parameters, which can be accessed through the VISUALISER.

The simulation tool

The last environmental exploration feature of the air bubble's TABLIER is its simulation module (Figure 7.5). An environmental calculator—AMBICALC—with a cubic design, is available to perform several modeling calculations; allowing, for example, prediction and evaluation of the effects of, say, an oil spill in the estuary, or an odorous release from the wastewater treatment plant. These simulations can be performed in an independent external application, LIVE SKETCH (Nobre and Câmara, 1995), which is an end-user tool for interactive spatial simulation, based on graphical sketching (Figure 7.6). The AMBICALC performs other useful calculations, such as unit conversions, constant values and scientific formulations.

The Environmental Glossary tool

An Environmental Glossary was designed, so that users may have a full description of the environmental terms. This hypermedia structured glossary is accessible through the TABLIER or through any hypertext or hyper-object link that the user selects.

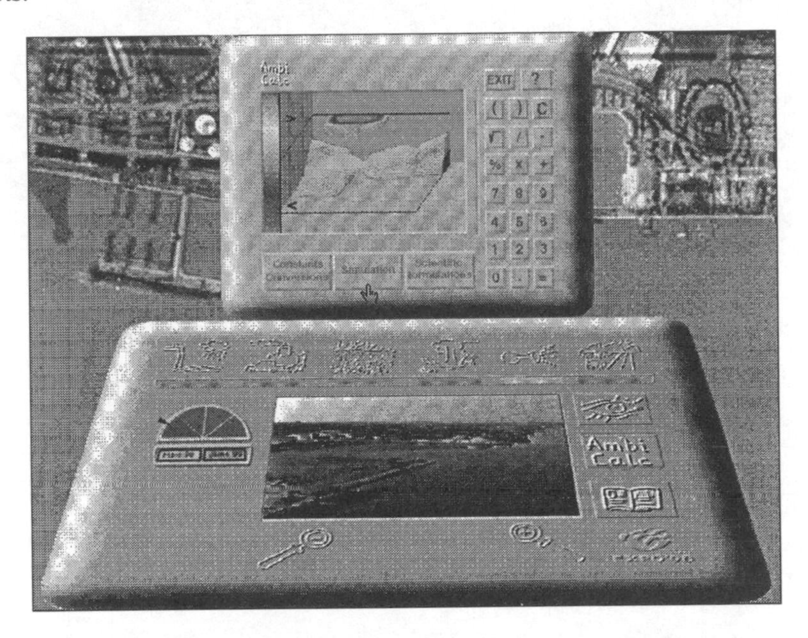

Figure 7.5 AMBICALC—a simulation and environmental calculator tool.

In this glossary, users may find textual descriptions of environmentally related terms, enhanced with photographs, video clips and other sources that may enrich the description.

The CD-ROM also includes a sequential way to access the information, as an alternative to the interactive exploration described above. This option follows, as closely as possible, the story of the EXPO'98 project implementation process. It starts with a description of the site before implementation, emphasizing the most critical environmental problems, followed by the construction and exhibition phases.

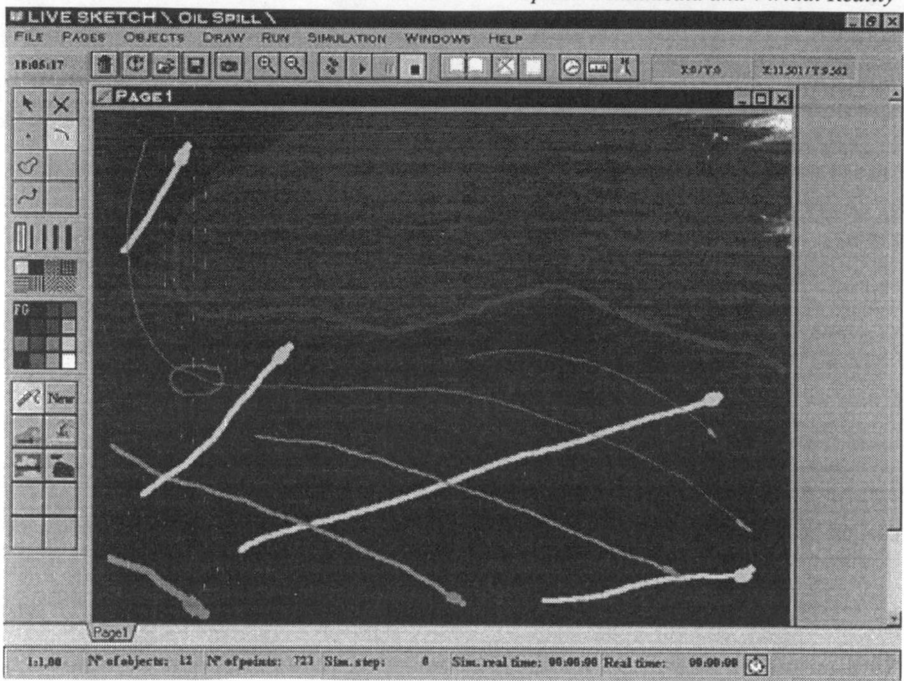

Figure 7.6 LIVE SKETCH—an interactive spatial simulation tool based on graphical sketching.

The user may be a passenger on a guided tour, on which the air bubble will be blown along by the prevailing winds, to give an aerial overview of the existing site before the project; carried through the river or estuary flow, to realize water and sediments pollution; or taken up by the rain and infiltrated into the soil, to identify the sources of groundwater and soil contamination.

7.4 FUTURE DEVELOPMENTS

Future developments will include updating of the information that will be produced for the area after 1998. Also, a version of the CD-ROM that is implemented in HTML will eventually be launched on the WWW to complement the existing site (www.expo98.pt/ambiente). As the HTML and VRML (Virtual Reality Modeling Language) formats evolve , this WWW distributed version of the system will be adapted to the most familiar and most often used standard.

As the scenes are composed of still photographs, it is possible to provide the users with real photographs of the site. The interaction, although limited in these so-called 'movies', gives the user an impression of actually 'looking around'. These scenes will provide the user with an extremely realistic and interactive view of the EXPO'98 site.

Finally, future developments associated with environmental and planning multimedia applications will rely on two major issues: the development of data retrieval tools and the availability of distributed tools for visualization, analysis and collaboration.

As far as data retrieval is concerned, the main issues are the expected availability of high-resolution satellite data, as well as data from remote sensors, associated with the growing number of national spatial data infrastructures around the world.

The availability of modular tools on the WWW, benefiting from advances in interoperability and customizable according to users' needs, will contribute to the creation of distributed components that are useful for specific environmental and planning activities, exploring the visualization, analysis and collaborative features provided by multimedia technologies.

ACKNOWLEDGMENTS

This work was partially supported by JNICT/DGA under research contract no. PEAM/C/TAI/267/93. We would like to acknowledge the special collaboration of Parque EXPO S.A., which provided part of the information that is being used in the development of the CD-ROM. We also would like to acknowledge the collaboration of Edmundo Nobre, for his work on LIVE SKETCH.

REFERENCES

ARNOLD, U. and ORLOB, G. T., 1989, Decision support for estuarial water quality management, *Journal of Water Resources, Planning and Management*, **115**(6), 775–92.

BLAT, J., DELGADO, A., RUIZ, M. and SEGUI, J. M., 1995, Designing multimedia GIS for territorial planning: the ParcBIT case, *Environment and Planning B: Planning and Design*, **22**, 665–78.

CÂMARA, A. S., 1989, A decision support system for the Tejo Estuary, in *International Seminar: Water Quality Assessment and Management*, Lisbon, 17-19 May, PGIRH/T Press, pp. 135–47.

CÂMARA, A. S., FERREIRA, F. C. and FIALHO, J. E., 1994, Pictorial modelling of dynamic systems, *System Dynamics Review*, **10**(4), 361–73.

CARVER, S., 1995, Position paper for the I-17 Initiative, NCGIA Specialist Meeting on Collaborative Spatial Decision-Making Initiative 17, Santa Barbara, CA, September, http://www.ncgia.ucsb.edu:80/research/i17/I17_home.html.

CONKLIN, J., 1985, Hypertext: an introduction and survey, *IEEE Computing*, **20**(9), 17–40.

CRAWFORD, C., 1990, Lessons from computer game design, in LAUREL, B. (Ed.), *The Art of Human–Computer Interface Design*, Reading, MA: Addison-Wesley, pp. 103–11.

FEDRA, K., 1993, Interactive environmental software: integration, simulation and visualisation, RR–90–10, International Institute for Applied Systems Analysis, Laxenburg, Austria.

FERNANDES, J. P., FONSECA, A., PEREIRA, L., FARIA, A., FIGUEIRA, H., HENRIQUES, I., GARÇÃO, R. and CÂMARA, A., 1997, Visualization and interaction tools for aerial photograph mosaics, *Computers & Geosciences, Special Issue on Exploratory Cartographic Visualization*, **23**(4), 465–74.

FERRAZ DE ABREU, P. and CHITO, B., 1997, Current challenges in environmental impact assessment evaluation in Portugal, and the role of new information technologies: the case of S. João da Talha's incinerator for solid urban waste, in REIS MACHADO, J. and AHERN, J. (Ed.), *Environmental Challenges in an Expanding Urban World and the Role of Emerging Technologies*, Lisbon: National Center for Geographical Information (CNIG), pp. 1–11.

FERREIRA, F., 1995, Digital video applied to air pollution emission's monitoring and modelling, in *First Conference on Spatial Multimedia and Virtual Reality* (I CSMVR), Museu da Água, Lisbon, 18–20 October, pp. 21–7.

FONSECA, A., GOUVEIA, C., CÂMARA, A. S. and SILVA, J. P., 1995, Environmental impact assessment using multimedia spatial information systems, *Environment and Planning B: Planning and Design*, **22**, 637–48.

GEODAN, 1997, European spatial metadata infrastructure, Report submitted to the European Commission, Luxembourg.

GIBBS, S. and TSICHRITZIS, D., 1995, *Multimedia Programming: Objects, Environments and Frameworks*, New York: ACM Press.

GORDON, T., KARACAPILIDIS, N. I. and VOß, H., 1996, ZENO—A mediation system for spatial planning, in BUSBACH, U., KERR, D. and SIKKEL, K. (Eds.), *CSCW and the Web—Proceedings of the 5th ERCIM/W4G Workshop*, Sankt Augustin, Germany, Arbeitspapiere der GMD 984, pp. 55–61.

GOUVEIA, C., 1996, Augmenting public participation with information technology in a Portuguese environmental assessment context, Submitted to the Department of Urban Studies and Planning in partial fulfillment of the requirements for the degree of Master in City Planning at the Massachusetts Institute of Technology, USA.

GUARISO, G. and WERTHNER, H., 1989, *Environmental Decision Support Systems*, Ellis Horwood Series in Computers and their Applications, Chichester: Ellis Horwood.

HENRIQUES, R. G., 1996, The Portuguese National Network of Geographical Information (SNIG Network), in *Second Joint European Conference and Exhibition on Geographical Information*, Barcelona, http://snig.cnig.pt.

JOFFE, B. and WRIGHT, W., 1989, SIMCITY: thematic mapping + city management simulation = an entertaining, interactive gaming tool, in *Proceedings of GIS/LIS '89,* Orlando, FA, November, pp. 591–600.

KHOSHAFIAN, S. and BAKER, A., 1996, *Multimedia and Imaging Databases*, San Francisco: Morgan Kaufmann.

LANG, L., 1992, GIS comes to life, *Computer Graphics World*, October, 27–36.

LAUREL, B., 1990, Interface agents, in LAUREL, B. (Ed.), *The Art of Human-Computer Interface Design*, Reading, MA: Addison-Wesley, pp. 124–30.

LOUCKS, D. P., KINDLER, J. and FEDRA, K., 1985, Interactive water resources modelling and model use: an overview, *Water Resources Research*, **21**(2), 95–102.

NIELSEN, J., 1995, *Multimedia and HypertextCthe Internet and Beyond*, Boston: Academic Press.

NOBRE, E. and CÂMARA, A., 1995, Spatial simulation by sketch, in *First Conference on Spatial Multimedia and Virtual Reality* (I CSMVR), Museu da Água, Lisbon, 18-20 October, pp. 99–104.

RAPER, J., 1997, Progress in spatial multimedia, in CRAGLIA, M. and COUCLELIS, H. (Eds.), *Geographic Information Research: Bridging the Atlantic*, London: Taylor & Francis, pp. 525–43.

RHIND, D. W., ARMSTRONG, P. and OPENSHAW, S., 1988, The Domesday machine: a nationwide GIS, *Geographical Journal*, **154**, 56–68.

ROMÃO, T., CÂMARA, A., MOLENDIJK, M. and SCHOLTEN, H., 1995, Coastal management with aerial photograph based mosaics, in *First Conference on Spatial Multimedia and Virtual Reality* (I CSMVR), Museu da Água, Lisbon, 18-20 October, pp. 150–8.

SHIFFER, M., 1995a, Issues on collaborative spatial decision-support in city planning contexts, NCGIA Specialist Meeting on Collaborative Spatial Decision-Making, Initiative 17, Santa Barbara, CA, September, http://www.ncgia.ucsb.edu:80/research/i17/I-17_home.html.

SHIFFER, M., 1995b, Interactive multimedia planning support: moving from stand-alone systems to World Wide Web, *Environment and Planning B: Planning and Design*, **22**, 649–64.

SHIFFER, M. J., 1993, Augmenting geographic information with collaborative multimedia technologies, in *Proceedings of AUTO CARTO 11*, Minneapolis, pp. 367–76.

SHNEIDERMAN, B., 1998, *Designing the User Interface: Strategies for Effective Human-Computer Interaction*, Reading, MA: Addison-Wesley.

SMITH, T. and FREW, J., 1995, Alexandria digital library, *Communications of the ACM*, **38**(4), 61–2.

STOTTS, P. D. and FURUTA, R., 1991, Hypertext 2000: databases or documents? *Electronic Publishing*, **4**(2), June, 119–21.

THURING, M., HANNEMANN, J. and HAAKE, J. M., 1995, Hypermedia and cognition: designing for comprehension, *Communications of the ACM*, **38(8)**, August, 57–66.

CHAPTER EIGHT

Evoking the visualization experience in computer-assisted geographic education

David DiBiase

Deasy GeoGraphics Laboratory, Department of Geography, The Pennsylvania State University, University Park, PA 16802, USA

8.1 THE VISUALIZATION EXPERIENCE

Traditionally, US university-based cartographic design studios such as the one that I direct have served researchers primarily by preparing graphics for publication in journals and books. By extending the design expertise developed for the print medium into animated and interactive media, my colleagues and I have attempted in recent years to contribute also to researchers' effective use of interactive graphics for data exploration, hypothesis generation and testing—activities that distinguish *visualization* from graphic communication.

More recently, I have turned my attention from the needs of researchers to the needs of educators and, especially, of students. For me, visualization by undergraduate students is a more compelling problem than visualization by research specialists, because the latter have expertise which makes them less likely to be confused or misled by problems posed by graphic data representations.

The project described in this chapter involves authoring a set of interactive software modules which supplement a physical geography textbook. The question I wish to raise here is how our experience with visualization in scientific research can be applied to the design of educational software and to improve geographic education. I would like to entertain the notion that although the production of educational software is necessarily an act of communication, in certain circumstances the software can, and should, evoke in students what I will call the *visualization experience.*

For MacEachren (1994), geographic visualization is characterized by the manipulation of graphic data representations by individuals who seek to construct new knowledge. I will define the 'visualization experience' as the moment at which learners (be they experts or novices), working alone or in unsupervised small groups, become aware of an unanticipated question as a result of an interaction with graphic data. Such an awareness is a necessary condition for the cognitive processes involved in the transformation of graphic data into information, as well as the social processes involved in the construction of knowledge.

Instructional multimedia can, in principle, provide the interactive environment needed to foster the visualization experience. As noted above, however, instructional visualization systems must accomodate novice users. Further, while we can expect experts to be highly motivated, we cannot assume the same of undergraduate students. After setting the context for this case study I will describe the strategies that my software development team and I have adopted to compensate for inadequate expertise and motivation and to foster the visualization experience in computer-assisted geographic education.

8.2 THE NEW MEDIA LABS PROJECT

Public concern about the quality of higher education in the USA is widespread. One very tangible way in which Penn State and other US universities are responding is by providing instructors and students with greater access to networked computing and telecommunications technologies for classroom instruction and individual study. Unfortunately, the effort and expense involved in producing quality educational software that takes advantage of these technologies is daunting. For example, in a feature article on computer-assisted learning, *Byte* magazine reported that, on average, 228 hours are required to create one hour of computer-based training (Reinhardt, 1995). Distance education expert Michael Moore (1992) has estimated that 'it can take a team of five to ten people a full year to design the equivalent of a one semester three hour course'. Not surprisingly, many instructors who were once enthusiastic about the prospect of creating their own customized courseware are now seeking off-the-shelf software, or the assistance of specialized software development teams (DeLoughry, 1993).

The Deasy GeoGraphics Laboratory has been active in educational software development since 1992. With internal funds provided by our Associate Dean for Education, we began by developing interactive software for use in classroom presentations of an introductory earth science course (Krygier *et al.*, 1997). In 1995 we were awarded a research and development contract by the textbook publishing firm John Wiley & Sons, Inc., to design and produce computer-based learning modules to supplement the introductory college text *Physical Geography of the Global Environment* (de Blij and Muller, 1996). The project resulted in a set of six interactive software modules called 'New Media Labs' which instructors can assign to students as laboratory exercises or study aids. We used Macromedia Director authoring software to produce the New Media Labs for use under both the Windows and Macintosh operating systems. The general objectives defined for the product were: (1) to develop instructional material that is effective in promoting quality physical geography education for entry-level college students; (2) to provide material that builds upon and extends, rather than merely reiterates, material presented in the text; and (3) to bolster sales of the book. Our larger project is to help realize the potential of interactive technologies to improve the quality of education in geography and related subjects.

Publishers such as Wiley invest cautiously in electronic supplements to their traditional print products. Recent market figures suggest that caution is appropriate: Only 4 percent of US CD-ROM development firms were profitable in 1994. Of the approximately 2000 multimedia CD-ROM titles available in the USA in December 1994, 400 had sold fewer than ten copies. Most of those that had sold well were games (Swisher, 1995). Even when publishers give away educational software as a free supplement to a textbook, it is used by less than one-third of its recipients (Vernon, 1993). Fortunately, publishing firms such as Wiley seem committed 'to taking responsible risks to address current classroom needs and explore future directions' (John Wiley & Sons, 1994).

8.3 TOPIC SELECTION

Given the understandable caution of our sponsor in investing in a risky product like the New Media Labs, it is no surprise that support for project staff and equipment was modest. The scope of the project enabled me to design only about half the modules that would be needed to provide a comprehensive set of laboratory exercises for the entire text. In collaboration with one of the authors of the text and Wiley's Geography Editor, I selected six topics that we expect will be of greatest interest to the target audience, while still representing the broad scope of the text. These include the Köppen climate classification system, global vegetation regions, climate dynamics, the greenhouse effect, enhanced ozone depletion, and volcanism. To provide coherence among this limited set of exercises, I adopted climate as a general theme.

8.4 INSTRUCTIONAL OBJECTIVES AND THEORIES OF LEARNING

Products such as the New Media Labs need to be reasonably successful in the marketplace if the publishing industry is to justify a continuing commitment to the development of innovative educational materials. The success or failure of such products depends on both the perceived quality of the software and the effectiveness of product marketing and promotion. In this university–industry collaboration, our primary role was to maximize the quality of the product. Our quality assurance strategy included reviews and applications of current research in learning theory, curriculum design for interactive educational materials, and the physical science content of the modules.

Instructional designers argue that defining explicit learning objectives is a crucial early step in developing software for computer-assisted learning (Squires and McDougall, 1994). I proposed a set of objectives for each module and distributed these proposals to the authors, editor and referees for review. Similarities among the objectives led me to consider the selected topics to be of two kinds: (1) specific techniques (such as the global climate and vegetation classification schemes) that need to be mastered by students before they can progress to more advanced course work; and 92) complex environmental issues with social implications (such as the greenhouse effect and enhanced ozone depletion), about which introductory-level students should be sufficiently familiar to be able to think critically.

Equally important to defining learning objectives is to adopt an underlying theory of learning that supports the stated objectives. Two predominant theories of learning are the *behaviorist* and the *constructivist* approaches. The behaviorist approach is exemplified by *drill and practice* software, in which students respond to programmed sequences of well-defined tasks (such as matching or true/false questions), with the software providing reinforcement through positive feedback. In contrast, the constructivist approach assumes that computer-assisted learning will be most effective if it provides open-ended tasks and simulation exercises, which require students to play a more active role in hypothesis generation and problem-solving. I chose to adopt both approaches, since each seemed well suited to one of our two categories of learning objectives. In what follows, I will illustrate the distinction between the approaches by describing features of two New Media Labs modules. I will also discuss the relative merits of each approach in evoking the visualization experience.

8.5 IMPLEMENTATIONS OF THE BEHAVIORIST APPROACH

In a review of applications of computer-assisted learning in geography, Unwin (1991) observes that although 'computers are being used as a routine teaching resource in very many departments of geography', many university instructors 'have a very limited view of how [computer-assisted learning] can be used'. He blames this view in part on 'an understandable rejection of ... a very mechanistic, "programmed learning" approach to the delivery and testing of factual information'. The bad reputation of so-called 'drill and practice' software is acknowledged by Weyh and Crook (1988), who go on, however, to provide a compelling defense of the approach. They draw a distinction between familiarity with a topic and mastery of it, arguing that certain techniques (such as equation-writing and problem-solving in their field, chemistry) 'must be mastered by students if they are to succeed in the subject'. Drill and practice software, they argue, can be effective in developing such mastery, provided that the software is well designed and implemented.

Introductory physical geography, at least as the subject is presented in *Physical Geography of the Global Environment,* does not require mastery of techniques comparable to an introductory chemistry course. However, such techniques are not entirely absent either. The Köppen climate classification system, for example, has been presented in textbooks for many years as a means of relating local temperature and precipitation observations to global-scale atmospheric processes that give rise to regions of climate similarity. Some authors and instructors continue to believe that the skills involved in classifying local climates by reading bivariate 'climographs' and applying the complex criteria of the Köppen system are useful for students who wish to progress to more advanced physical geography course work. Similarly, classification systems for vegetation regions and soil orders are common techniques for helping students to cope with the complexity of the environment.

The New Media Labs include two drill and practice modules designed to develop mastery of the Köppen system and a global vegetation classification. We assume that students have access to the *Physical Geography of the Global Environment* text, and that they will have read sections pertinent to each exercise before (or during) their interactions with the software. Like all the modules in the series, the Köppen laboratory consists of several sections, each of which is accessible to the user at any time by selecting an item in the Sections menu. Students can move from any 'page' of a module to any other by clicking the '?' button, which causes an interactive table of contents to appear. The introductory section of the Köppen module states the objectives of the module and provides background information on the Köppen system. The second section—classification—helps students to learn the temperature and precipitation criteria that define the climate categories, as well as the one-, two- and three-letter codes used to denote the categories. Students are presented with a textual description of a climate category, which they identify by clicking the button labeled with the correct climate code. The software responds to correct and incorrect matches with appropriate textual feedback. Students may call up a world map of the distribution of the climate region described in the text at any time by clicking a button labeled 'Where is it?' The order of presentation of the descriptions is randomized, so that students will not experience identical task sequences when they reuse the software.

The third section of the Köppen module introduces climographs as graphic representations of local temperature and precipitation data. The students' task is to identify the climate category that characterizes each climograph by clicking the appropriate climate code button (Figure 8.1). Forty-eight climographs have been prepared for the software; 24 (one for each category) are selected for each new use of the section, and their order is randomized. The location of each local climate on a world map can be displayed by clicking the climograph. Textual descriptions of each category can be called up in a window by selecting the category in a Notes menu. Correct, incorrect, and near-miss responses (when a student clicks a button in the correct major climate category, but not the correct subdivision) are acknowledged by the software with textual feedback. Besides developing students' mastery of the Köppen system, this section builds skills in interpreting quantitative information displayed in graphs.

Finally, a fourth section called 'climate map' emphasizes that the main value of the Köppen system is that it allows us to map the distribution of generalized climate regions at a global scale. After a description of the exercise, students are presented with a world map on which the Köppen regions are outlined, but not colored in. Below the map is an array of buttons labeled with climate letter codes. Students are instructed to 'paint' the climate regions by dragging a category color into its corresponding region (Figure 8.2). Correct matches cause the region to be painted; incorrect matches result in textual feedback.

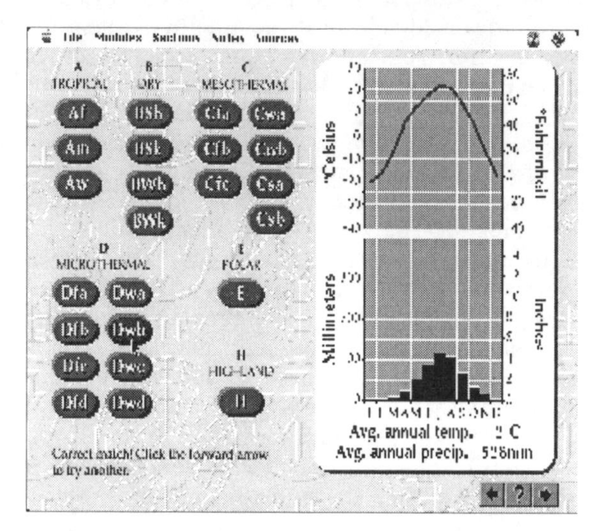

Figure 8.1 The climograph matching section of the New Media Labs Köppen climate classification module. Students identify the climate type that characterizes the data represented in the climograph by clicking on the appropriate climate code button. A map showing the location of the temperature and precipitation observations can be displayed by clicking on the climograph. © 1996 John Wiley & Sons, Inc. Used with permission.

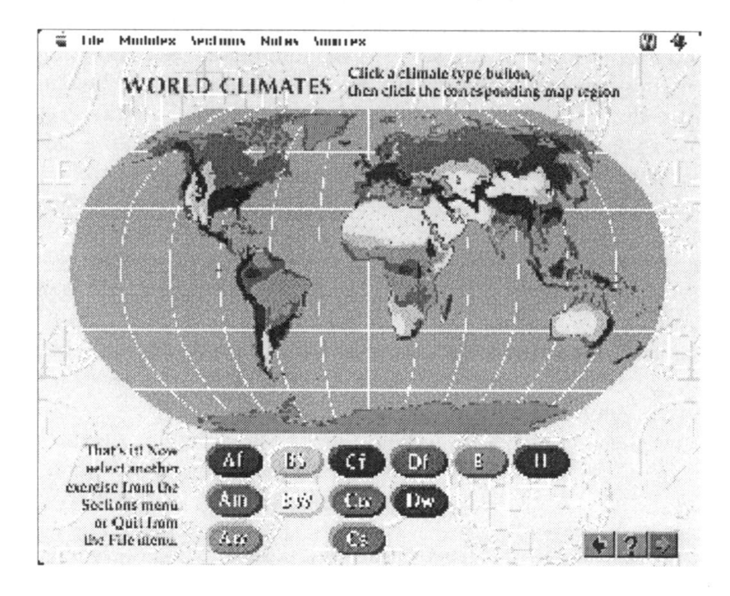

Figure 8.2 The climate map section of the New Media Labs Köppen climate classification module. Students drag colors from the climate code legend to the corresponding map region to color the map. © 1996 John Wiley & Sons, Inc. Used with permission.

The Köppen and Vegetation Regions modules of the New Media Labs series embody all the characteristics of successful drill and practice software identified by Weyh and Crook (1988). The modules require students to enter problem solutions on screen, they involve short answers (mouse clicks), feedback on student performance is provided immediately after each input, help is available to the student at any time (via the Notes menu), and navigation through the program is straightforward and consistent. Competitiveness is not exploited to leverage student motivation, since score-keeping, top ten lists and other arcade game hooks have been observed be less effective for females than for males (Caftori, 1994).

Except for the final section, only rudimentary visualization tasks are supported in the Köppen module. Maps are available to provide locational context for verbal descriptions and climographs in sections 2 and 3. The fourth section, in which students are asked to 'paint' a world map of Köppen climate regions, is intended to help students construct a mental image of climate distribution. In general, however, visualization plays only a minor role in these drill and practice exercises.

8.6 IMPLEMENTATIONS OF THE CONSTRUCTIVIST APPROACH

Four of the six New Media Labs modules are designed to help students learn about complex environmental issues—including climate dynamics, the greenhouse effect, enhanced ozone depletion, and volcanism—rather than to develop mastery over specific skills or techniques. In defining instructional objectives for these modules, I sought to identify the two or three most salient points about each topic that students would benefit most from understanding. If this seems too modest a goal for these modules, consider that a recent international survey of 25 000 people's knowledge of science issues suggests that 79 percent of the world's adults believe that the greenhouse effect is caused by ozone holes (Kierman, 1995).

The module on climate dynamics, for example, while providing a reasonably thorough overview of the subject, is designed to stress just three fundamental points. First, it seeks to provide students with the clearest possible sense of the distinction between climate *variability* and climate *change*. Second, it constructs a greatly simplified model of the Earth's energy balance that identifies the three variables of the climatic system that contribute most to long-term change: solar radiation, albedo and the greenhouse effect. Finally, it allows students to manipulate these variables and view resulting changes in global average surface temperature. Despite (or perhaps because of) its simplicity, the model appears to be effective in prompting students 'to question the reasons behind the facts generated by the computer model and to develop a "feel" for the principles involved' (Chatterton, 1985; cited in Squires and McDougall, 1994).

The climate dynamics module consists of three sections. As with other modules, the introductory section defines the objectives of the module and describes the other sections. The second section focuses on the distinction between climate variability and climate change. To provide students with a first-hand sense of the difficulty in discerning trends in short-term temperature fluctuations, the software provides a dynamic map of global surface temperature anomalies (relative to mean temperatures for a 1950-1980 reference period) from 1900 to 1993, accompanied by a line graph of the same data (Figure 8.3). Instead of an ordinary map animation that students are able only to watch, the module provides slider-controlled *flipbook animations* that enable students to control the rate of change of the sequence, to focus on subsets of the time series and to 'play' the sequence in reverse chronological order. Additionally, students can select an option to reorder the sequence from the year with the greatest negative global average temperature anomaly to the year with the greatest positive anomaly. This *reexpression* option may reveal regional patterns of cooling or warming that are obscured in sequences that are ordered chronologically (DiBiase *et al.*, 1992).

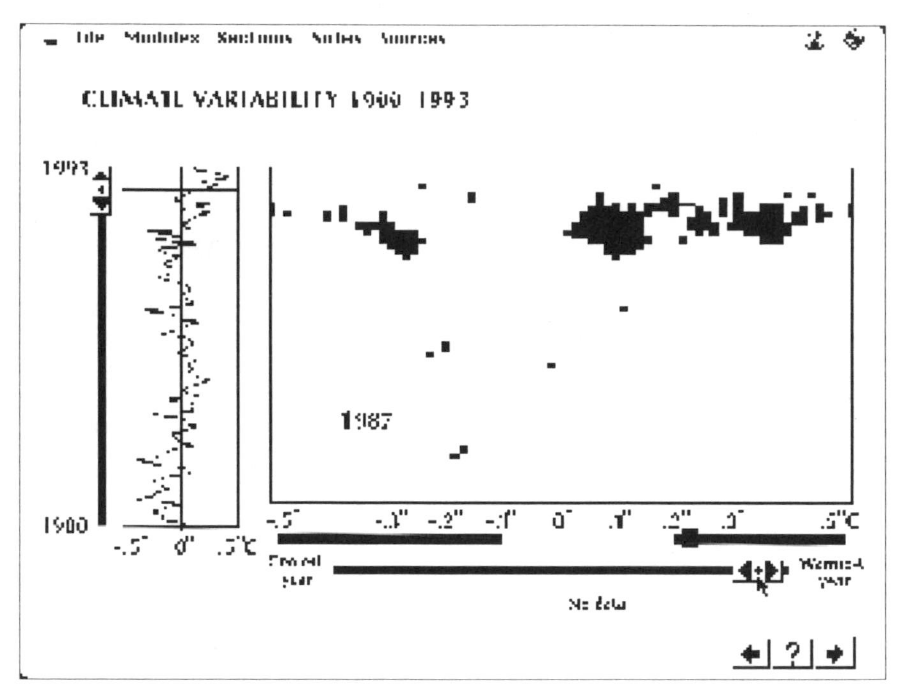

Figure 8.3 'Flipbook ' animation of global average temperature anomalies, 1990–1993. Students can 'play' the sequence in chronological order using the vertical slider, or, using the horizontal slider, from the year with the highest global average anomaly to the year with the lowest. © 1996 John Wiley & Sons, Inc. Used with permission.

The third section uses a predictive model of the Earth's energy balance to highlight the crucial variables of the climatic system that account for long-term climate change. The module employs a *sequential model construction* approach in which variables are introduced one by one, allowing each to be discussed as it appears. This approach also provides a means of disclosing details about the workings of the predictive model, building expertise that students need to critique the assumptions and simplifications on which the model depends. At each step of model construction, *hypertext links* are provided for students who prefer to explore the topic tangentially rather than linearly. For example, the screen that introduces solar radiation provides links to information on sun spots and Milankovitch theory. Finally, the module culminates in an *interactive model* of the Earth's energy balance (Figure 8.4). The model enables students to manipulate solar radiation, albedo and greenhouse effect values, and to observe the resulting change in global average surface air temperature. Students (and instructors) will be able to use the model to simulate basic problems in earth science, such as the 'faint young sun' problem.

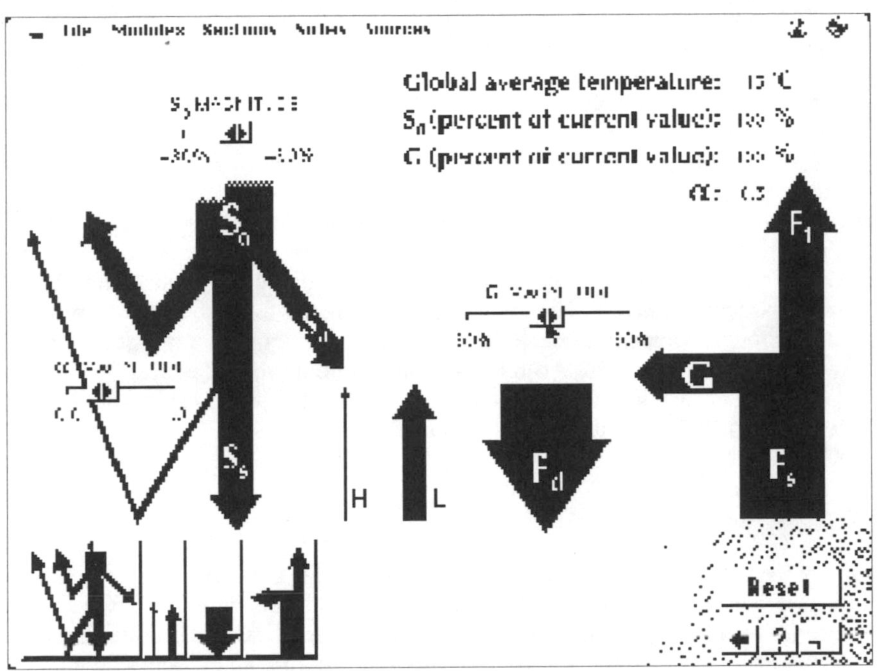

Figure 8.4 The interactive model of the global energy balance. Students can modify values of solar radiation, albedo and the greenhouse effect, and view resulting changes in global average surface air temperature. © 1996 John Wiley & Sons, Inc. Used with permission.

In contrast to the behaviorist drill and practice modules described in the preceding section, the four New Media Labs modules designed from a constructivist perspective accommodate several strategies for evoking the visualization experience in students. Flipbook animations, reexpression and interactive models enable students to make substantive changes in the representations that are presented. Sequential model construction and hypertext links provide opportunities for students to develop the expertise that they need to evaluate the material critically. Although specific learning objectives have guided the design of the modules, students are not required to commit particular information to memory or to solve predetermined problems. Rather, they are intended to support the open-ended exploration of evidence characteristic of the visualization experience.

8.7 DISCUSSION

I began this essay with the assertion that in certain circumstances educational software can, and should, evoke in students a visualization experience. The project described here suggests that visualization can be fostered in computer-assisted learning, particularly in software designed from a constructivist perspective. For meaningful visualization to occur, I expect that inadequate student expertise and motivation must be assumed and accounted for (McGuinness, 1994). The New Media Labs include concise textual summaries that build on the subject knowledge that students gain from lectures and textbook readings. Strategies such as sequential model construction and hypertext links bolster student expertise by providing convenient access to important details.

I assume that the motivation of students who use the New Media Labs will be enhanced in two ways. First, the Labs attempt to engage student interest by revealing the innately fascinating character of the environmental data, concepts and issues represented in the software. Instructional design strategies such as flipbook animation, reexpression and interactivity can be employed to arouse and channel students' curiosity. Second, I recognize that it remains for instructors to motivate students through compelling teaching and imaginative assignments that involve the software. Furthermore, while visualization and computer-based education share the goal of fostering insight, only instructors can provide the judgment required to help students interpret the quality of their insights.

8.8 EPILOG

Our contract with the publisher included a licensing agreement by which Penn State is authorized to provide New Media Labs to students without charge through its campus computing facilities. In the two years since I first drafted this chapter, at least 500 Penn State students have used the software in conjunction with an introductory physical geography course. Two hundred and forty-seven students completed surveys in which they were asked to evaluate the product. Only 9 percent reported previous experience with computer-assisted learning. Among other questions, the survey asked 'If the material in New Media Labs was also available in a book, would you prefer to study the computer-based material or the printed material? Why?' Sixty-two percent of students expected to prefer the computer-based format. Advantages identified by these students in response to this open-ended question included the software's interactivity and visualization capabilities. Many of the students who would have preferred print media complained that they had difficulty gaining access to the university's crowded computing facilities. Fifty-two percent of students surveyed in 1996 stated that they would have been able to use the CD-ROM version of New Media Labs (which was not available to them) on their own personal computers; 62 percent claimed to have suitable personal equipment in 1997. This result is consistent with a campus-wide survey (Moore, 1996), which found that two-thirds of Penn State undergraduates own multimedia-capable personal computers by the time they graduate. The results of national surveys (e.g. Green, 1996) indicate that while rates of student ownership of personal computers varies greatly among institutions and programs, access to computing hardware is posing less and less of an obstacle.

Despite its successful use at Penn State, the New Media Labs were anything but a commercial success. The CD was offered as an expensive optional enhancement to the text, and few instructors adopted it. Fortunately, the product showed enough promise that the Deasy Lab has since been commissioned to develop more elaborate multimedia products for human geography, world regional geography and physical geology.

Publishers' attitudes about multimedia software have changed dramatically over the past two years. The same editors who used to speak of software products as 'ancillaries' now refer to the textbooks themselves as the 'print component' of multimedia products. World Wide Web sites and CD-ROMs are now standard features of the educational materials offered to instructors of large-enrollment university courses. Combined with the rapidly expanding hardware infrastructure noted above, competition among publishers has led to greater opportunities in educational software development.

Developers' attitudes about learning theories are also changing. As Hutchings *et al*. (1992, p. 172) have concluded, 'the nature of human learning (and of computer use in general) is such that prescriptive guidelines have limited utility'. Ideological adherence to any particular theory of learning has come under fire by some who argue that integrative approaches are needed. Plowman (1996), for example, points out that highly interactive, unstructured tasks meant to foster active learning are often responsible for disorientation and frustration among students. She suggests that the linear structure of narratives—an anathema to strict constructivists— may help users to overcome such difficulties. The notion of a 'cognitive apprenticeship' (Collins *et al*., 1989) provides an intriguing model for developers who wish to integrate the advantages of behaviorist and constructivist perspectives. The cognitive apprenticeship 'ignores the usual distinctions between academic and vocational education, its objective being to initiate the novice into a community of expert practice' (Berryman, 1993, p. 4). In a forthcoming product entitled Geosciences in Action, my colleagues and I implement the cognitive apprenticeship concept as a set of 'virtual internships', in which students use interactive graphics as they are guided through applied problems confronted by professional geologists.

Finally, distance education has changed from a topic of discussion to an action item for many universities, government agencies and private firms over the past two years. Administrators recognize that increasing populations of adult learners are not well served by the traditional resident model of instructional delivery. As we adapt to this new educational context, the ability to evoke the visualization experience in computer-assisted learning will become a necessity.

ACKNOWLEDGMENTS

I am grateful for the opportunity to have particpated in the GISDATA Specialist Meeting on GIS & Multimedia (Rostock, Germany, 1994) and in the First Conference on Spatial Multimedia and Virtual Reality (Lisbon, Portugal, 1995). I wish to thank Barbara Buttenfield for her recommendation, as well as the National Center for Geographic Information Analysis, the National Science Foundation, and the European Science Foundation for their support. Janine Acee and Mark Wherley of the Deasy GeoGraphics Laboratory made crucial contributions to the production of the New Media Labs. I thank Frank Lyman, former Geography Editor of John Wiley & Sons, Inc., for his role in realizing the project. Special thanks go to Cindy Brewer, for her insightful suggestions on this chapter.

REFERENCES

BERRYMAN, S. E., 1993, Designing effective learning environments: cognitive apprenticeship model, Institute on Education and the Economy, Brief #1, www.tc.comumbia.edu/~iee/briefs/BRIEF1.HTM.

CAFTORI, N., 1994, Educational effectiveness of computer software, *T.H.E. Journal,* August, 62-5.

CHATTERTON, J. L., 1985, Evaluating CAL in the classroom, in REID, I. and RUSHTON, J. (Eds.), *Teachers, Computers and the Classroom*, Manchester: Manchester University Press, pp. 88-95.

COLLINS, A., BROWN, J. S. and NEWMAN, S., 1989, Cognitive apprenticeship: teaching the craft of reading, writing, and mathematics, in RESNICK, L. B. (Ed.), *Knowing, Learning and Instruction: Essays in Honor of Robert Glaser*, Hillsdale, NJ: Lawrence Erlbaum.

DE BLIJ, H. J. and MULLER, P. O., 1996, *Physical Geography of the Global Environment*, 2nd Edn, New York: John Wiley.

DELOUGHRY, T. J., 1993, Academic-computing directors give low priority to developing instructional software, survey says, *The Chronicle of Higher Education*, 20 October, A26-7.

DIBIASE, D., MACEACHREN, A. M., KRYGIER, J. B. and REEVES, C., 1992, Animation and the role of map design in scientific visualization, *Cartography and Geographic Information Systems*, **19**(4), 201-14, 265-6.

GREEN, K. C., 1996, *Campus Computing, 1995*, Encino, CA: Campus Computing.

HUTCHINGS, G. A., HALL, W., BRIGGS, J., HAMMOND, N. V., KIBBY, M. R., MCKNIGHT, C. and RILEY, D., 1992, Authoring and evaluation of hypermedia for education, *Computers in Education*, **18**(1–3), 171–7.

JACKSON, G. A., 1990, Evaluating learning technology, *Journal of Higher Education*, **61**(3), 294-311.

KIERMAN, V., 1995, Survey plumbs depths of international ignorance, *New Scientist*, 29 April, 7.

KRYGIER, J. B., REEVES, C., DIBIASE, D. and CUPP, J., 1997, Design, implementation, and evaluation of multimedia resources for geography and earth science education, *Journal of Geography in Higher Education*, **21**(1), 17–38.

MACEACHREN, A. M., 1994, Visualization in modern cartography: setting the agenda, in MACEACHREN, A. M. and TAYLOR, D. R. F. (Eds.), *Visualization in Modern Cartography*, Oxford: Pergamon, pp. 1–12.

MCGUINNESS, C., 1994, Expert/novice use of visualization tools, in MACEACHREN, A. M. and TAYLOR, D. R. F. (Eds.), *Visualization in Modern Cartography*, Oxford: Pergamon, pp. 185–99.

MOORE, B. L., 1996, Survey on the computer experience of Penn State students,

Center for Academic Computing, the Pennsylvania State University.
MOORE, M. G., 1992, Take time to design, *The American Journal of Distance Education*, **6**(2), 1–2.
PLOWMAN, L., 1996, Designing interactive media for schools: a review based on contextual observation, *Information Design Journal*, **8**(3), 258–66.
REINHARDT, A., 1995, New ways to learn, *Byte,* March, 51–72.
SQUIRES, D. and MCDOUGALL, A., 1994, *Choosing and Using Educational Software: a Teachers' Guide*, London: The Falmer Press.
SWISHER, K., 1995, The tarnish on the silvery disc: CD-ROM pioneers are finding that being first isn't always safe, *Washington Post*, 29 May–4 June, 21.
UNWIN, D. J., 1991, Using computers to help students learn: computer assisted learning in geography, *Area*, **23**(1), 25–34.
VERNON, R. F., 1993, What *really* happens to complimentary textbook software? A case study in software utilization, *Journal of Computer-Based Education*, **20**(2), 35–8.
WEYH, J. A. and CROOK, J. R., 1988, CAI drill and practice: is it really that bad? *Academic Computing*, May/June, 32–54.
JOHN WILEY & SONS, INC., 1994, *1994 Annual Report*, p. 8.

CHAPTER NINE

Spatial simulation by sketching

Edmundo M. N. Nobre and António S. Câmara
Environmental Systems Analysis Group, New University of Lisbon,
2825 Monte de Caparica, Portugal

9.1 INTRODUCTION

Drawing is one of the most intuitive and innate methods of human communication. Indeed, Man drew before he could write or even speak. A sketch is a special kind of sparse drawing, usually produced quickly and in a personal manner. It consists of a set of graphical objects drawn on a background. These objects may be recalled from memory or constructed by the imagination (Scrivener *et al.*, 1994).

Sketching has been used to define graphical objects in a user interface. Efforts have been made to represent through sketching the relationships and connections, the performance and attributes, and even movement of objects (Lansdown, 1994; Scrivener *et al.*, 1994).

Graphical user interfaces representing either artificial concepts or natural objects and phenomena are based on a set of graphical objects. These objects may be defined by sketching. Sketched objects are strokes defined through freehand drawing, using a pen-based input device.

Our proposal is that sketching may be used, in addition, to trigger behaviour and interaction rules between the objects and, thus, to enable their evolution through time.

To accomplish this goal, we have developed an approach named Live Sketch. This methodology is based on the implementation of visual evolution rules that are dictated by sketching. Our work is based on previous contributions on interactive sketching (Landay and Myers, 1995), pictorial simulation (Câmara *et al.*, 1993), cellular automata (CA) (Toffoli and Margolus, 1987) and programming by demonstration (Smith *et al.*, 1994).

9.2 IMPLEMENTATION OF LIVE SKETCH

To implement Live Sketch, four main issues were studied: the definition of visual evolution rules; algorithmic implementation of those rules; numerical decoding; and computer implementation.

9.2.1 Visual evolution rules

Graphical objects are the basic units of Live Sketch. These objects are defined by sketching through their visual variables: shape, size, colour and position. The evolution of these objects in space and time is expressed by the transition of their visual variables from one initial state to a final one.

Sketched objects belong to specific user-defined groups. These object groups are identified by a set of transition rules. Transition rules include behaviour and interaction rules. Objects can move, expand, retract, decay, multiply or just change colour (Figure 9.1). They can also be attracted, repulsed or absorbed, or intersect one another (Figure 9.2).

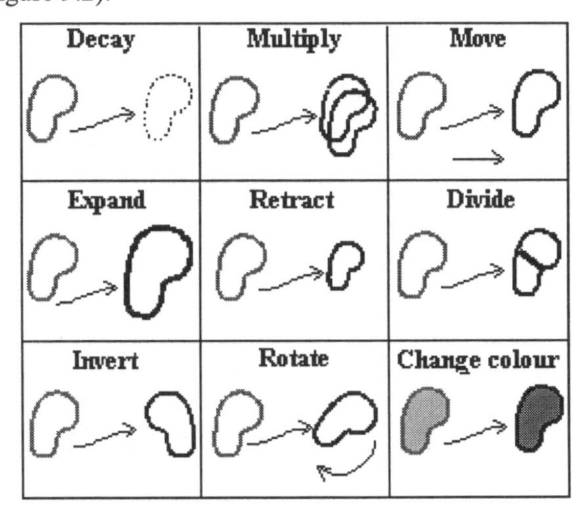

Figure 9.1 Behaviour rules.

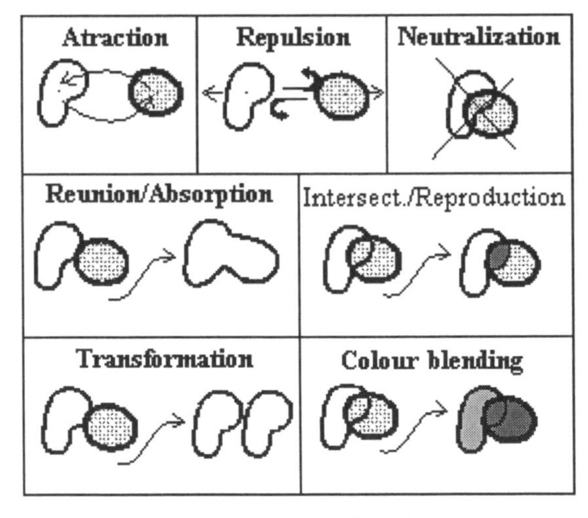

Figure 9.2 Interaction rules.

Sketches may also be used to activate interaction and behaviour transition rules such as the ones presented in Figures 9.1 and 9.2. In this case, sketches are used as demonstrations to be followed in the actual simulation, an idea proposed by Smith *et al.* (1994).

9.2.2 Implementation of algorithm

To implement visual evolution rules, cellular automata (CA) formulations were developed. These formulations are ideally suited for spatial simulation. Thus, cell values and transition rules had to be defined.

Sketched objects are represented in the cellular automata formulation as a collection of cells that define the shape, size and position of the object (Figure 9.3). The cell values define their colour, which may be changed through simple transition rules.

The fundamental problem relates to changing the object's shape (which, in turn, conditions the size and position of the object). To solve this problem, we applied the vector field concept. In these fields, each cell contains information on orientation and intensity. This information is inferred from sketched arrows (Figure 9.4), which are also used to trigger the shape deformation through time.

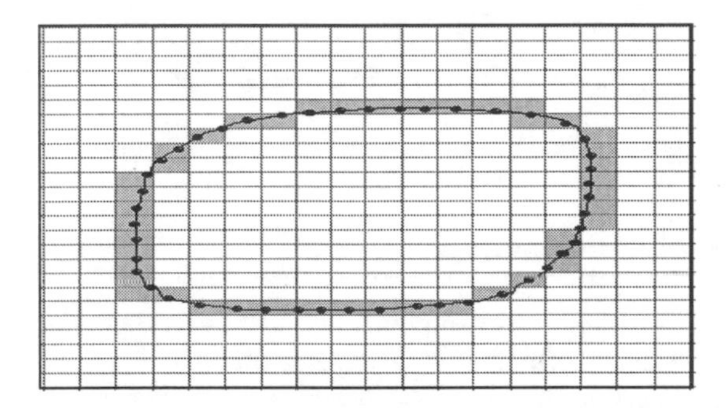

Figure 9.3 Sketch an object/create a CA object.

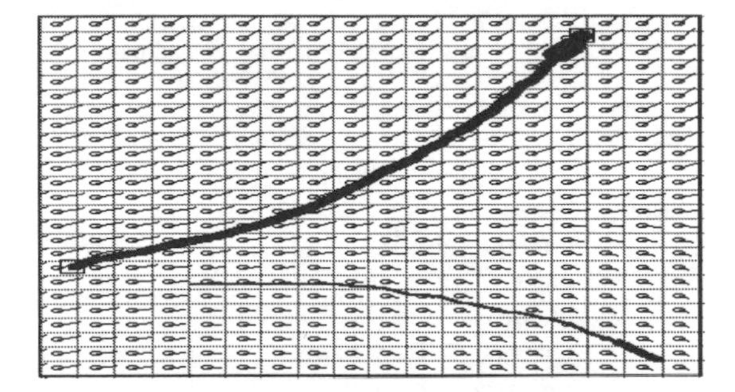

Figure 9.4 Sketch an arrow/create a vector field.

9.2.3 Numerical decoding

Spatial information has multidimensional properties that can be extracted or expressed directly through image analysis. Shape, size, colour, position and pattern are the basic kinds of graphical information related to an image. Some of these properties can be directly decoded into numerical quantities.

Shape is directly related to the object volume, area and perimeter and also to fractal dimensions. Colour may be associated with relevant variables such as density, temperature or memory size. These basic numerical variables may be primitives in numerical expressions, enabling the definition of other variables.

Once graphical variables change in time, the related numerical variables will also obviously change with them. Conversely, numerical changes may contribute to modifications in the graphical representations.

An example of the interaction between graphical and numerical representations is given in the simulation of a growing cell. In this example, the cell will grow (expand) until a certain dimension is reached (the area value). It will then split (divide) into two (see Figure 9.5).

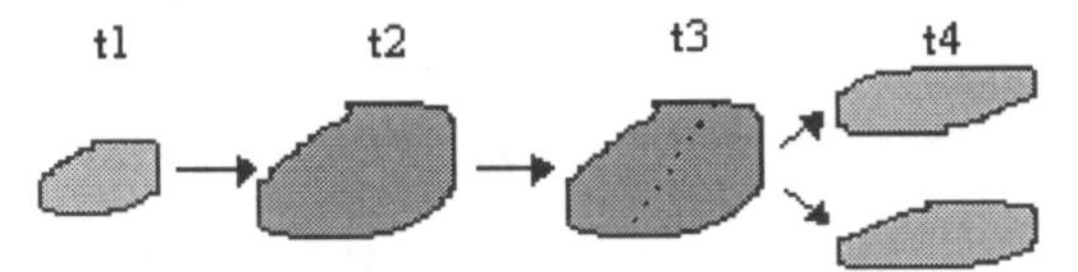

If area(celll) > xx then divide(celll) into 2

Figure 9.5 Numerical operations with an object.

9.2.4 Computer implementation

A prototype of Live Sketch was implemented in Visual Basic©. The user interface of Live Sketch follows principles that are common to many successful drawing applications, as shown in the Live Sketch Toolbox (Figure 9.6).
The correspondence of the drawing tools to the Live Sketch procedures is synthesised in Table 9.1.

Table 9.1 Basic tools and related procedures.

	Object	Properties	Relations
Drawing tools	Select, draw, cut, copy, paste	Fill	Arrows
Procedure	Create, redefine, destroy, duplicate	Expansion, retraction, decay, multiply	Attraction, repulsion, movement

Live Sketch also includes a sketch interpreter. The user may draw a single point, an open line, a closed line (irregular polygon) or an arrow. In the interpreter, all of these single points and lines are converted to arrays of points with a regular distance between them, creating the basis for the application of the cellular automata formulations described above.

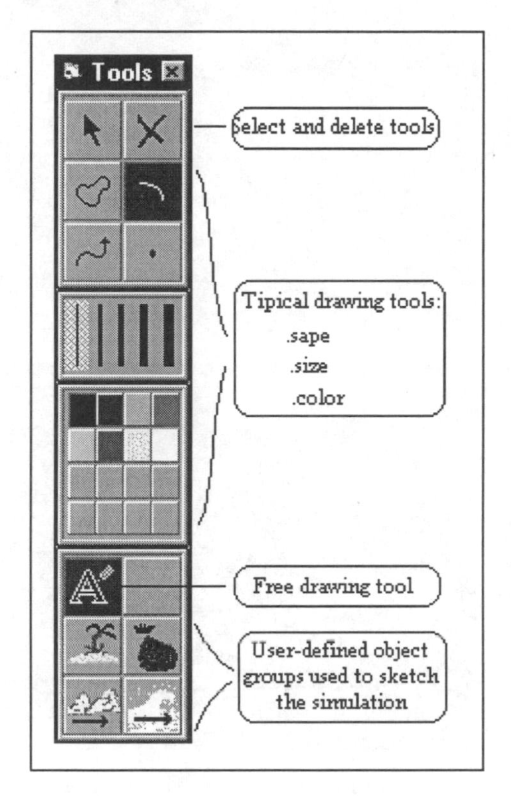

Figure 9.6 The Live Sketch Toolbox.

9.3 APPLICATIONS

Live sketching can be readily applied to spatial simulation problems, such as the drift simulation of pollution plumes produced in a oil spill accident. Other examples might include coastline evolution through time, forest fires and air pollution.

To illustrate Live Sketch, we have developed a simple oil spill model. An aerial photograph was used as a background (Figure 9.6) and we considered that the movement of floating oil in the sea was the result of the combined influences of current, wind speed and orientation (International Tanker Owners Pollution Federation, 1986).

Sketching those influences as arrows (currents; wind) with varying orientations (arrow direction) and intensities (arrow thickness), and drawing obstacles such as land, one can simulate the evolution of the oil spill. Figure 9.7 illustrates the application (1 = currents, 2 = wind, 3 = land, 4 = oil).

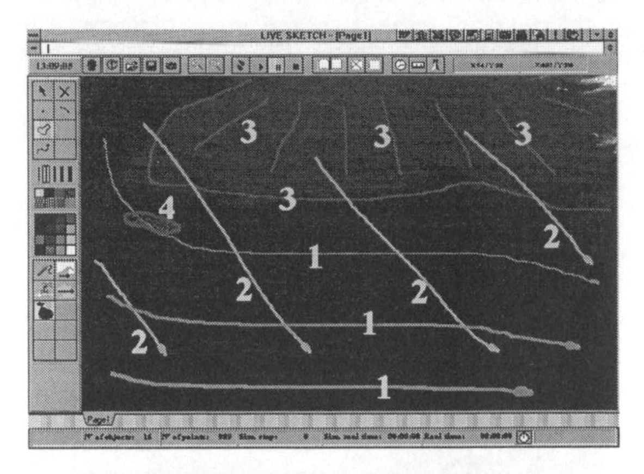

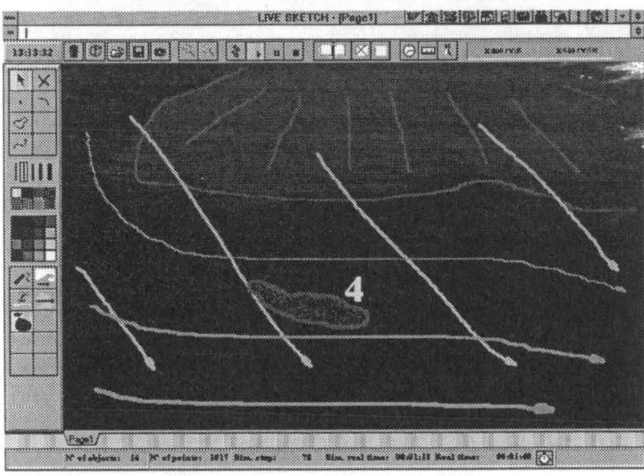

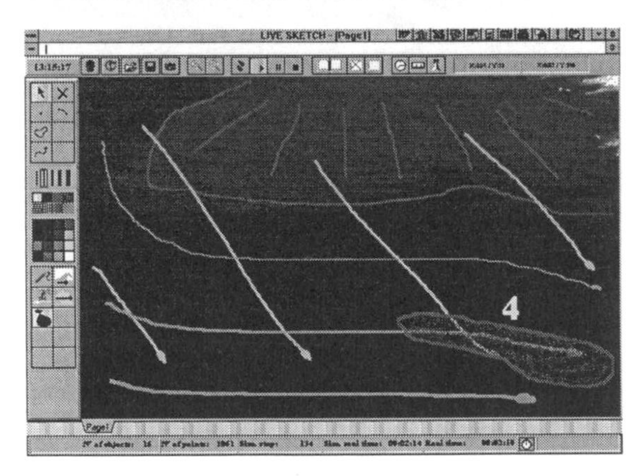

Figure 9.7 An oil spill example.

Non-calibrated live sketching is appropriate whenever specific data is not available or precise results are not needed. But sketching may also be used as an interface to rigorously calibrated numerical models, using cellular automata formulations.

Live Sketch may also be applied in many other fields, such as education, art, architecture, engineering and user interface design. Indeed, whenever graphical objects change through time, Live Sketch may be used to trigger, interact with and display their evolution.

9.4 FUTURE DEVELOPMENTS

Several developments are foreseen for the proposed approach. Some will result from the application of Live Sketch to different situations. For these, usability tests are also planned, to evaluate and improve Live Sketch performance when dealing with these problems.

Two other developments the use of dynamic background images, and the expansion of live sketching to network environments are planned by the authors.

In the first development the intention is to use background video with two goals: video images may be used to calibrate spatial simulation efforts; and sketching may be used to create interactive experiences within the video itself. At the present time, we are mainly interested in ecological movies and attempts to use sketching tools to quickly simulate their evolution.

The expansion of Live Sketch to the Internet environment is also planned. A network-based dynamic sketching tool could be used to enable communication among different experts around the world, in the real-time simulation of catastrophes.

9.5 CONCLUSIONS

We have presented an approach based on sketching to define graphical objects and simulate their evolution. This approach, named Live Sketch, is based on the consideration of visual variables of objects, and of the behaviour and interaction rules that govern their values through time.

It has been shown that modelling of the change of shape is the key step in interactive sketching. To simulate the evolution of shape using sketching, a cellular automata formulation has been proposed. Sketching has been used in this formulation to define cell values and trigger the application of transition rules.

To illustrate the proposed approach, an application to an oil spill problem was presented. Future developments relate to the insights gained with the application of Live Sketch to other problems, the use of video in the background and the expansion of the approach to network environments.

ACKNOWLEDGEMENTS

This work was supported by *Junta Nacional de Investigação Científica e*

10.2 DEFINING AGENTS

Many definitions have been given for agents. In fact, the term has been used to 'describe everything from a word processor Help system to mobile code that can roam networks to our bidding' (Wayner, 1995a). Several definitions of agents exist, given by different researchers, each involving the characteristics that are most valuable to them (Franklin and Graesser, 1996). We have chosen Wooldridge and Jennings' (1995) interpretation, because although it is a very general definition it includes key features that will provide an initial classification of agents: 'An agent is a self contained problem solving entity implemented in hardware, software or a mixture of the two' that should include the following properties:

- *Autonomy*. Agents should be able to execute their problem-solving tasks without direct intervention from humans or other agents. They should, to some degree, control their own actions and internal state.
- *Social ability*. Agents should be able to interact, when they see fit, with other agents and humans, either to complete their problem-solving or to help other agents.
- *Responsiveness*. Agents should hold knowledge on their environment and be able to respond to any changes that may take place in this environment.
- *Proactiveness*. Agents should not only act in response to their environment, but they should be able to take advantage of fortuitous opportunities to achieve their designated goals. An agent should be able to modify its behaviour in response to stimuli.

Agents can be classified in several ways (for some classifications, see Wooldridge and Jennings, 1995; Franklin and Graesser, 1996). Because this chapter is concerned with the application of agents to spatial issues, we will only discuss the types of agent that will be useful in this area of expertise: mobile agents, reactive agents, cognitive agents and interface agents.

10.2.1 Mobile agents

Mobile agents are programs that may be dispatched from a client computer and transported to a remote server computer for execution (Harrison *et al.*, 1995). It has been suggested that mobile agents offer an important new method of performing transactions and information retrieval in networks. Mobile (itinerant) agent architectures can be considered as an extension to client/server computing, in which the client creates additional processes that are sent through the network to search and retrieve the required information. These processes are defined according to the types of interaction that they perform (Chess *et al.*, 1995):

- *Information dispersal/retrieval*. Simple interactions based on an ask/receive paradigm between an itinerant agent and a static agent.
- *Collaborative*. A more complicated type of interaction in which there is a single clearly defined goal. The agents not only ask for and receive information, but they also evaluate and compromise according to their

preferences.

- *Procurement*. Very complex interactions governed by an auction protocol, in which the agents' goals and resources are hidden from other agents.

The same authors describe in detail one scenario for a Travel Reservation System using itinerant agents. Wayner (1995c) presents mobile agent architectures by defining the underlying technology and the roles given to different players in the architecture: roaming mobile agents, host processes that control and enable the agents' activity, and resources to be used by the agents. Security issues, such as protecting the host computer from foreign agents attacks or bugs, or preventing other users from taking control of arriving agents, are a very important concern in mobile agents.

10.2.2 Deliberative (cognitive) agents

Deliberative agents are agents that contain an explicitly represented, symbolic model of the world and in which decisions are made via symbolic reasoning (Wooldridge, 1995).They have a memory, they build plans of action, they can be selfish or co-operative, and they exchange complex messages (Ferrand, 1995).

Wooldridge (1995) presents several efforts within the symbolic AI community to construct deliberative agents. However, these architectures usually suffer from the following problems:

- *The transduction problem*. Translating the real world into an accurate and adequate symbolic description, in time for that description to be useful.
- *The representation/reasoning problem*. How to symbolically represent information about complex real-world entities and processes, and how to enable agents to reason with this information in time for the results to be useful.

The existing algorithms for symbol manipulation cannot guarantee termination with useful results in an acceptable fixed time bound, which is essential in agent systems. This led researchers to look into reactive agent architectures.

10.2.3 Reactive agents

Wooldridge (1995) defines reactive agents in a negative fashion (as opposed to deliberative agents). Reactive agents do not include any kind of central symbolic model and do not use complex symbolic reasoning. They are generally very simple structures that act by reacting directly to changes that take place in their environment. There is no global control in the system: global behaviour emerges from the local reactive actions of each agent. In section 10.3, a survey of multi-reactive-agent systems related to spatial issues is presented. Current developments also show researchers creating hybrid systems using agents that mix cognitive with reactive characteristics (Wooldridge and Jennings, 1995).

10.2.4 Interface agents

Interface agents are semi-intelligent, semi-autonomous systems that assist users when dealing with one or more computer applications (Kozierok and Maes, 1993). The metaphor used is that of a personal assistant collaborating with the user in his or her work environment. Research into this type of agent has been under way for some time. Maes (1994b) argues that there are several approaches to building interface agents and defends a Machine Learning approach. She states that, under certain conditions, an agent can 'program itself'. The agent is given some initial knowledge and evolves by learning 'behaviour' from the user and from other similar agents. At the MIT Media Laboratory, several interface agents have been built to help users to accomplish tasks such as electronic mail handling, the scheduling of meetings, the filtering of news, and the selection of entertainment. For detailed descriptions of the characteristics of these agents, see Maes (1994b).

10.3 SPATIAL AGENTS

A spatial agent is an autonomous agent that can reason over representations of space. Having a goal with spatial characteristics to fulfil, it must be able to access and/or handle spatial information, reason on it and execute the appropriate tasks. A spatial agent will make spatial concepts computable.

Agents with a spatial awareness are a relatively old concept. In Minsky (1986), space is described as 'just a society of nearness relations between places' and the global geography of a space as 'nothing more than hints about which pairs of points lie near one another'. Minsky also states that several layers of agents build the maps inside our brains, each layer being composed of agents that are responsible for regions of the space and whose function is to detect which other agents are the nearest to them. According to Franklin and Graesser (1996), Minsky's agents may not be considered autonomous agents, as far as the definition that we are considering is concerned. Minsky's agents are fine-grained agents that form a mind as a whole.

The concept of Cellular Automata is also very close to that of agents: 'Cellular Automata are mathematical models for systems in which many simple components act together to produce complicated patterns of behaviour' (Wolfram, 1994). Cellular automata are composed of a regular lattice of sites, each site taking on k possible values. The current site is called a cell. The values are updated in discrete time steps according to a rule that depends on the value of the sites situated in some neighbourhood around the cell. Cellular Automata include several of the features that characterise agents, although on a simpler scale. The basic unit is the cell. The automaton evolves as the cells change state according to what is happening in their neighbourhood (environment). From the local changes in the state of cells, a pattern (goal) emerges. Ferrand (1995) quotes work at LAMA-IGA (France) related to the simulation of accessibility of space through the road network from any point. The first attempt to find a solution involved Cellular Automata, but the researchers later moved on to Multi-Reactive-Agent Systems (MRAS). This solution can naturally deal with any type of complex diffusion process, such as paths crossing above bridges. It can also easily handle a simulation in continuous time, making agents act in different threads of execution. In MRAS, intelligence emerges from the interaction of multiple simple entities, acting on the basis of direct reactions to stimuli (Ferrand, 1995). Reactive agents are thus very close to the Cellular Automata paradigm, as well as to connectionism. In fact, existing architectures seem to blur the frontier between one concept and the other. Ferrand (1995) also quotes projects related to cartographic generalisation and Multi-criterion Spatial Decision Support where MRAS have been in use. The former aims to 'support the creation of maps starting from a set of data, taking into account a thematical focus and graphical constraints linked to the used symbols and the final resolution'. The latter concerns single-actor decision-making and spatial negotiation support (specifically, the choice of a path or an area for large-scale infrastructure, such as a road or electrical power line).

At the Santa Fe Institute, the Swarm simulation system was developed in order to provide researchers with a set of standardised simulation tools (Minar et al., 1996). The formalism is that of a collection of independent agents interacting via discrete events. In addition to being containers of agents, swarms can themselves be agents and an agent can also be a swarm. Simulations can, thus, involve several levels of complexity.

Also, Toomey *et al.* (1994) describe a project that aims to implement software agents to help the dissemination of remote sensing data, and present an architecture in which agents communicate with each other in order to locate, or, if necessary, produce, the information requested by the user. This agent-based Remote Terrestrial Sensing (RTS) data dissemination environment allows the user to specify the location of the desired imagery's geographical region (by drawing it directly on a world map) and to specify constraints on other image attributes. These constraints are introduced using generic RTS domain terms instead of database specific ones. The agents in the system are in charge of data sources. A Data Broker Agent leads the communication between agents. These are very different agents from those studied by Ferrand. However, they do retain a spatial reasoning component, as they have to be able to handle knowledge about the space being portrayed in the RTS images, as well as knowledge about their specific environment of distributed data sources.

10.3.1 Research agenda

We have just covered current research on agents with spatial characteristics. However, it is our opinion that the most interesting opportunities for research in spatial agents have yet to be addressed:

- *Location and retrieval of spatial information in large networks*. The growth of the Internet has made huge resources available to everyone. To control and assess these resources, it is necessary to create structures that help to locate and retrieve the right information. This problem becomes more intense with spatial information due to the large volumes of data manipulated and the necessity for spatial indexes. There is also an important concern about metadata standards for the publication of spatial data on the Internet. Different agents can be built to locate the necessary information for the user, and to filter the retrieved information by identifying the user's needs. Spatial data mining problems are covered in section 10.3.1.1.

- *Facilitate the handling of a GIS user interface*. This thread will use interface agents research to create agents that acquire knowledge about the tasks, habits and preferences of users. These agents should be able to execute tasks on the user's behalf or suggest actions to be taken. Another possible development is the dynamic re-formulation of the GIS user environment according to the user's evolving profile. GIS interface agents are discussed in section 10.3.1.2.
- *Implementation of improved spatial tasks*. The use of GIS becomes more difficult as new software releases become available and functionality expands. The large amount of information manipulated again has to be considered. Agent functionality in this area could include the generation of templates for executed tasks, the monitoring of these tasks, and the creation of intelligent spatial data that evolves as new information becomes locally or globally available. Spatial tasks are covered in section 10.3.1.3.
- *Creating interfaces between GIS and specific software packages*. This is a consequence of the growing number of fields of application for GIS. The use of spatial models, statistical packages, generalisation algorithms or any other application that adds functionality to GIS is currently very common. Unfortunately, the interaction between GIS and these packages is practically non-existent and has to be provided manually by the user. Agents cannot only provide for this interface but they can also add power to the packages through personalisation. This issue is discussed in section 10.3.1.4.

10.3.1.1 Spatial data mining

Agents performing spatial data mining, after receiving the specification of their search, should travel through the network looking for the data that the user needs. The best way to implement this kind of agent will, therefore, be to use mobile computing (see section 10.2.1). In this type of architecture, several issues must be considered:

- *Server (host) implementation*. A spatial information server available on the Internet (and/or the World Wide Web) must publish its information in such a way that agents will be able to find it and recognise its value to their user. Therefore, the structure and publication of the servers' metadata is as important as the data itself. In addition, it is necessary to define concepts common to the agents and the server (ontologies; see Gruber, 1993), creating a common vocabulary that will enable them to communicate and exchange information successfully. Current research into digital libraries is addressing these problems (*Communications of the ACM*, 1995; *IEEE Computer*, 1996).
- *Client implementation*. The agent should be able to take action on a complete as well as an incomplete specification of request. A user may know exactly what he wants and where it is held, or he or she may only have a general idea. That is why personalisation is important. The agent can trace back and try to find similar requests in previous situations. Also, if every personal searching agent acquires information about the resources utilised by its user, then communication and co-operation among agents may provide them with

valuable information on unknown resources. Using machine learning techniques, it is possible to build knowledge about which agents to ask for help when looking for a specific type of information. It is also possible to create group profiles so that agents will automatically be able to contact peers that belong to the same group or domain. This type of work has been addressed by the autonomous agents group at MIT Media Lab (Maes, 1994a).

- *Execution of request.* Once the right information or resource has been located, there are two ways in which to execute the requested tasks: send an agent to the remote resource location to request the needed service and retrieve the results; or have the remote server send the functionality and/or data so that the request can be executed locally. It is obvious that the right choice depends on the request being made (i.e., filtering, analysis, queries). However, the choice lies between the use of mobile agent systems or the development of client applications that could possibly be built using commercially available on-line mapping technology.

10.3.1.2 Improving GIS Interfaces

Interface agents will help the user in his or her daily work with the GIS. They will constantly learn the user preferences and profile and build knowledge about the way in which he or she works. As it learns, the agent will suggest automation of tasks to the user. The evolution of the rate of right suggestions will help the user to trust the agent. Slowly, the agent will start to execute tasks, on the user's behalf. The agent can also modify the user's environment according to his or her preferred tools or the commands that are executed most often. The user's environment will constantly and dynamically evolve. The Open Sesame! Assistant for the Macintosh provides this kind of functionality on the Finder application. It 'listens' to the user's commands and builds a profile based on what it has gathered. If a pattern is identified, the agent will offer to automate the task that the pattern refers to. The user's confidence is built up slowly, as the percentage of 'right' suggestions increases. The agent can also build knowledge through Programming by Demonstration (PBD) (Cypher, 1993). An experienced GIS user may be asked to 'show' a learning interface agent how it should execute sets of tasks in order to achieve a specific goal. The agent will be told when and where it should start to learn and when the learning process is complete. With the gathered information, it should then be capable of repeating the process when the right situation presents itself. One related experience is that of Campos *et al.* (1996), who describe a knowledge-based interface agent for ARC/INFO that receives and processes the user's requests in plain English. The agent takes this information and generates sequences of commands that ARC/INFO can understand. If the concepts known to the user are not confirmed by the ARC/INFO database, it interacts with him or her to clarify the misconception. After execution, the agent delivers and presents the results to the user.

10.3.1.3 Facilitating spatial tasks

This type of agent will be developed in order to integrate proactive spatial processes into complex GIS applications. Simulation environments or spatial decision support systems will be the best areas for application of this kind of agent. Depending on the complexity of these processes, these agents will either be reactive or deliberative (cognitive). The need for the manipulation of symbolic contextual information will be a major concern. The information available to the user will evolve autonomously as new data becomes available to the application or as new simulation cycles are initiated. One example of this type of application is the prototype of the MEGAAOT projects (Rodrigues *et al.*, 1998). An environmental planning application for spatial decision-making at the local level is based on a multi-agent control system. The spatial processes involved evolve dynamically as changes are issued into a land-use map. Dependencies on processes and data are dynamic links that may or may not be effective at one moment *t*. The environmental performance of a specific change at a specific time is stored as added experience for later use.

10.3.1.4 Connecting spatial systems

These agents will serve as interfaces between GIS and external packages, such as spatial models, statistical packages, generalisation algorithms. These agents will provide for an input interface for external commands, they will be responsible for communicating with the external package, and they will finally integrate the results into the GIS data model. It is our belief that this type of agent can range from a very simple process that sends some input to an application, to a (perhaps distributed) complex application that is transparently integrated with the GIS.

10.4 TOOLS FOR SPATIAL AGENT DEVELOPMENT

Several development tools have become available for the implementation of these agents. Next, we will evaluate some of these tools according to the properties for agent development and the necessities of spatial agents.

10.4.1 Properties for agent development

With the increasing popularity of agents, several development tools have emerged with the intention of making agent development easy and natural. There are also several existing tools that have been adopted by researchers and developers for the implementation of agents.

According to the various problem areas in which agents can be useful, and taking into account the agent characteristics mentioned in section 10.2, we can say that tools can be classified according to the following properties:

- *Modularity*. The ability to create agents that are capable of handling simple, well defined tasks. This provides agent reusability and enables the use of divide and conquer strategies[1] when dealing with complex problems, thus resulting in

[1] The use of the term ‘divide and conquer’ does not imply the use of static problem solving modules. These modules are obviously agents themselves, subject to the properties of proactiveness and therefore capable of evolving.

simpler programming and better overall system architecture. This is a fundamental property associated not only with agent development but with good code in general.

- *Interaction among agents.*[2] The interface provided for agent communication should be flexible and robust, as run-time communication is not known at design time. An agent must be prepared to communicate with any other agents, in a location- and time-independent fashion.
- *Access to distributed resources.*[2] An agent's access to data and knowledge, despite their location. In some cases the agent may even be able to change its location in order to better achieve its goals. This is related to the next property.
- *Mobility.*[2] In some domains, agents will have improved efficiency if they are capable of moving in a network. If this is a heterogeneous network, transportability of code is an issue. Agents must be able to change their location and continue to run without needing to recompile.
- *Knowledge manipulation.* The agent's ability to store knowledge, update the knowledge it holds ('learn') and plan the actions to be taken considering the changes. The chosen tool should have structures and mechanisms that provide for knowledge representation, machine learning and planning. In the perfect situation, the tool that would integrate these characteristics has fundamental concepts.

On the basis of this assessment of agent development characteristics, we will now review several tools that are currently used for agent development. We will be mainly concerned in classifying these tools with regard to their adequacy for the development of the types of spatial agents mentioned above.

10.4.2 Agent development tools

In this section we present a study of the characteristics of several widely distributed programming languages and development tools which are currently being used to build agents. The list is not comprehensive. The aim is to identify what makes a tool adequate for (spatial) agent development and not to provide a complete list of tools and languages.

10.4.2.1 New tools

Java The Java language was designed to meet the challenges of application development in the context of network-wide distributed environments (Gosling and McGilton, 1995). It is a very robust object-oriented language (with the exception of primitive data types, everything in Java is an object), as all memory management is executed by the interpreter, including automatic garbage collection. Being an object-oriented language, it satisfies the modularity requisite, therefore

[2] The issue of security of information across a network is not considered here, as it is not relevant to the discussion.

providing for inheritance, encapsulation and dynamic binding. There is no special feature for communication among agents. However, the messaging mechanisms typical of OO languages and the built-in capacity for multi-threading will facilitate the development of agent communication structures. It is an architecturally neutral language: this means that a program written in Java can be sent to and run on any system on which the Java interpreter and run-time system have been installed. The language is the same on any platform. However, it provides no knowledge-handling features or machine learning and planning functionality. Recently, several Java-based agent development libraries have been made available, providing the necessary functionality to add knowledge and machine learning capabilities to agents systems. Communication structures between agents are also widely provided (e.g. Bits & Pixels Intelligent Agent Library; see http://www.bitpix.com/business/main/bitpix.htm). Development tools for the creation of mobile Java agents are also becoming commercially available. One example is IBM's Aglets Workbench (Lange and Chang, 1996).

KSE The Knowledge Sharing Effort (KSE; see http://www.cs.umbc.edu/kse/) aims to 'develop techniques and a methodology for building large-scale knowledge bases which are shareable and reusable' (Finin *et al.*, 1994). KSE aims to provide building blocks for interaction and interoperation, which are the two characteristics that are found to be most desirable in agent systems (Finin *et al.*, 1997). Several tools have been developed in order to achieve this goal:

- *KQML (Knowledge Query and Manipulation Language)*. This is a message-format and message-handling protocol to support run-time knowledge-sharing among agents. It is therefore an interface for communication among agents. KQML focuses on an extensible set of performatives, the operations that agents use to communicate.

- *KIF (Knowledge Interchange Format)*. This is 'a formal language for the interchange of knowledge among disparate programs (written by different programmers, at different times, in different languages and so forth)' (ARPA KSE, 1995). KIF has not been built to be an internal representation for knowledge, but to serve as a common format for the interchange of knowledge between programs.

- *Ontolingua*. 'An ontology is an explicit specification of a conceptualisation' (Gruber, 1993). It describes the concepts and relationships that an agent or community of agents have in common. Ontolingua provides descriptions of ontologies in a form that is compatible with multiple representation languages. The syntax and semantics of Ontolingua definitions are based on KIF. Each ontology defines a set of classes, functions and object constants for a specific domain, including constraints for interpretation (Finin *et al.*, 1997).

The KSE tools provide for robustness and flexibility in communication among agents, even if written in different languages, by different people. They also enable representations of knowledge to be shared by several different systems. However, these are not tools to build agents, but to help them communicate and co-operate when trying to achieve their goals.

TCL (and Agent TCL) TCL stands for 'Tool Control Language' and is a simple scripting language for controlling and extending applications. Its popularity comes from being embeddable and extendible: 'Its interpreter is implemented as a library of C procedures that can easily be incorporated into applications, and each application can extend the core TCL features with additional commands specific to that application' (Ousterhout, 1993). Although TCL was created for a very different purpose, its internal structure, cost and reputation make it a good candidate for agent programming (Wayner, 1995c): TCL is an interpreted scripting language that encourages modular programs with many small tools; tools and their metaprograms pass around TCL scripts in the same way that computers may want to despatch mobile agents. TCL is not object-oriented, but its being interpreted and extendible makes it quite easy to add functionality to applications and to execute them in different machines.

Safe-TCL is an extension to TCL that allows for 'foreign' agents to run safely on a machine. These 'unsafe' agents will only be able to execute specific actions, those that will not endanger the environment in which they are running. Wayner (1995c) states that Safe-TCL contains all the hooks that allow it to read incoming mail and evaluate it as a script in a safe manner. This incoming script will execute without being able to thrash the host computer.

The qualities of TCL for agent programming have led to the development of a (mobile) agent system (Agent TCL; see Gray *et al.*, 1997). Agent TCL includes migration, message passing and graphical interface (through Tk) facilities. Security issues are considered through the use of Pretty Good Privacy for encryption and authentication. Safe-TCL enforces access restriction in the host computer.

Telescript 'Telescript is a set of technologies that provide foundation for electronic messaging, distributed processing and remote programming with communicators, computers, telephones and the networks that link them together' (Knaster, 1995). Included is a programming language with which it is possible to implement and customise powerful messaging systems. Messages in Telescript include agents that can execute tasks in a Telescript-aware network. According to Wayner (1994), Telescript Agents include 'built-in intelligence about how to interact with other systems', which is a key advantage if we realise that the language is as computationally powerful as C or BASIC. Of course, in order to have Telescript agents running on your machine and communicating among themselves, you must have installed a Telescript engine. Telescript is an object-oriented language that enables mobile agents in a telescript network to interact in a robust way and access data and resources at different locations. As Wayner (1994) says, they hold intelligence on the systems that they interface with. However, there are no mechanisms for knowledge handling, machine learning or planning. Recently, General Magic, the company responsible for the creation of Telescript, seems to be redirecting its agent development to the Java and ActiveX platforms (see http://www.genmagic.com).

10.4.2.2 Existing tools

Some of the already existing languages have been adopted for agent development thanks to their inherent capabilities. Some of these were even improved to better support agent development.

Lisp Lisp is one of the oldest higher-order languages in the computer world that survives through several dialects, each maintaining a part of the initial characteristics of the language. Recently, a committee of Lisp programmers approved the ANSI CLOS (Common Lisp Object System) standard. This standard defines a Dynamic Object-Oriented Language that allows for all the normal OO features plus some more. It provides automatic and efficient use of multiple inheritance, automatic memory management, method dispatching and unlimited scalability. There are even particular implementations that include extensions to the development environment for constructing complex knowledge-based systems. These extensions have been created to help construct applications, but they are not part of the language itself. Lisp has been considered as a good language for developing agents, because programs and data are stored in trees built from nested lists. A program can build a structure and then execute it. This becomes important in the implementation of learning agents that need to build new algorithms for automation of tasks. Also, some of the implementations of Lisp are interpreted, allowing for the development of transportable code (and perhaps mobile agents). At the MIT Media Lab, the Autonomous Agents group used Lisp to build their learning agents.

It is clear that the modularity property of our list is fulfilled by ANSI CLOS. The knowledge-handling property is partly provided by a body of extensions developed in specific implementations. There is, however, no support of network-based functionality.

SmalltalkAgents Being another object-oriented programming language (in which literally everything is an object) Smalltalk supports inheritance, class and instance behaviour, dynamic binding, messaging and garbage collection. Its OO nature provides the modularity property and the messaging for writing agents that communicate with each other. To improve its capacity for agent development, Quasar Knowledge Systems has developed SmalltalkAgents, a full-feature object-oriented authoring environment which includes tools for 'designing, developing and managing frameworks, classes and objects for use in the development of sophisticated components and agents' (Quasar Knowledge Systems, 1995). SmalltalkAgents includes: a visual, drag-and-drop rapid application development environment for user-interface layout and component design; native Windows components, multi-threading, and exception handling; easy integration with external (C, C++, Java, Visual Basic) code and data structures; and an extensive library of Smalltalk classes. According to Wayner (1995c), SmalltalkAgents is intended to make it simple for programmers to develop applications that run on a distributed network of machines with different CPUs. Quasar has created a low-level device independent language and all of SmalltalkAgents programs are effectively compiled into this language. In this way, their programs can be run on any machine that supports this low-level language without needing to recompile.

CLIPS CLIPS is 'a productive development and delivery expert system tool which provides a complete environment for the construction of rule and/or object based expert systems' (NASA, 1995). CLIPS supports three programming paradigms: rule-based (knowledge is represented as heuristics), object-oriented (complex systems are modelled as modular components) and procedural (capabilities similar to those provided by C, Pascal and Lisp). There are some current efforts to create extensions to CLIPS, adding functionality for producing agent systems. DYNACLIPS (Cengeloglu *et al.*, 1994) is a set of blackboard, dynamic knowledge exchange and agent tools for CLIPS. Agents communicate through the blackboard, sending and receiving facts, rules and commands. Another example of an agent tool based on CLIPS is AGENT_CLIPS (see http://users.aimnet.com/~yilsoft/software/agentclips/agentclips.html). The agents created in AGENT_CLIPS are rule-based agents which can scan newsgroups and web pages looking for information. They can also exchange knowledge with other agents (facts and rules) at runtime.

These are cases of development tools that integrate at its root the possibility of knowledge representation with object-oriented and rule-based programming. The OO features provide for modularity, and the fact that it was built primarily as an expert system development tool make it easier to solve Knowledge Engineering problems.

10.4.2.3 Why use only one language?

Because of the wide use that the term 'agent' is given today, there is no one language that can on its own shoulder the burden of agent development. As we have seen, each tool has its own qualities and the most complex agents may be built from getting all of these strong features together. However, some tools are quite complete for developing one specific type of agent. Some agent tools developers try to provide an environment that is as complete as possible for the kind of agent that they want to specialise in. The problem is always making a complete specification of the requirements for the system that we aim to build.

Table 10.1 Classification of development tools according to the relevant properties

	JAVA	Telescript	Tcl	SmallTalk Agents	KSE	Lisp	CLIPS
Modularity	*	*	+	*	-	+	*
Robustness and flexibility in communication among agents	#	*	#	*	*	-	#
Mobility	+	*	#	-	-	+	-
Distributed data and resources	*	*	*	*	*	+	*
Knowledge	#	-	-	-	*	+	*

Notes: *The tool supports the property completely. +The tool partially supports the property. #Support is provided through extensions to the tool.

Table 10.1 is an attempt to classify the studied development tools according to the properties specified in section 10.4.1. Satisfying the modularity property are all the Object-Oriented languages: Java, Telescript, SmalltalkAgents and CLIPS. TCL is classified as partially supporting the property because, although it is not OO, it encourages the development of modular code. Only some implementations of Lisp are OO, so we classify it, too, as only providing some support of modularity. Robust and flexible communication among agents is provided by Telescript, SmalltalkAgents and the KSE tools. In fact, the major aim for the development of the KSE tools was to provide sharing of information between agents. Extensions to Java, Tcl and CLIPS also enable communication among agents. The only language that explicitly implements functions for mobile agents is Telescript (but only inside a Telescript-aware network), although TCL extensions (Safe-TCL and Agent TCL) also provide some facilities for mobility. Access to distributed data and resources is given by any interpreted language. Any program written in an interpreted language (such as some implementations of Lisp) will be able to access information in hosts where its interpreter resides. SmalltalkAgents and Java are not interpreted, but their code is generated into low-level device independent languages, so their programs may run wherever those languages are understood. The KSE tools are not used to create agents, but they make heterogeneous agents understand each other.

We conclude that Telescript and TCL (especially Agent TCL) are the best choices for implementing mobile agents. However, this can also be done with Java. Interface agents will be best implemented when using Lisp, as this language enables the creation of new algorithms resulting from the learning of the agent. Lisp will also be a good choice for deliberative agents, as lists can be used to store symbolic knowledge. Extensions to CLIPS can be used for the development of either interface and deliberative agents. Reactive agents do not make major constraints on languages, so we will not consider any best choices. The KSE tools are fundamental in making heterogeneous agents communicate.

10.4.2.4 How to implement spatial agents?

To implement spatial data mining agents is more complex than creating mobile agents. It is necessary to choose tools for implementing the server, for creating metadata knowledge, for the implementation of the mobile searching agents, for interface agents (specification of requests) and for communication among agents. Therefore, this architecture will probably need mobile agents, interface agents and deliberative agents.

When implementing GIS interface agents, there is a need to integrate GIS code with a learning algorithm which is probably implemented in Lisp or another language with similar properties. Monitoring of events in the GIS and creating new commands for automating tasks will have to be transparently executed by the GIS but controlled by the agent code.

Agents facilitating spatial tasks will either be deliberative or reactive. In the deliberative case, the concern with the manipulation of symbolic knowledge suggests the use of Lisp or DYNACLIPS, although intelligent agent libraries for Java can also be used. Reactive agents leave most possibilities open.

Finally, connection agents are an open issue. Because of the amazing amount of possibilities for their implementation, we will not suggest any agent architectures or tools. This will have to be decided according to the problems to be solved.

This analysis of spatial architectures and tools is not intended to be complete. The objective was to look into some possibilities and choose possible paths to be taken. In the next section, we will present a prototype of a GIS interface agent that has been developed for the Smallworld GIS drawing tool. It is a very simple prototype, but its implementation was very useful for the authors because it raised some additional questions about the development of spatial agents.

10.5 A PROTOTYPE OF A GIS INTERFACE AGENT

10.5.1 Agent architecture

In collaboration with SmallworldWide, Cambridge, an interface agent architecture for the drawing and plotting tool of the Smallworld GIS was developed. Smallworld GIS is an Object-Oriented GIS which was developed in an OO language called Magik. This language is part of Smallworld and most of the code that forms the GIS is available for use to make changes. It is, therefore, very easy and natural to include monitoring structures, and to add additional features to the code. Because this initial agent does not include any reasoning, it was decided that Magik was sufficient to build it. The agent architecture is composed of two types of agent (Figure 10.1):

- The agent controller (Figure 10.2)—the agent that monitors the GIS.
- The task agents—each task agent automates a specific tool of the GIS and/or helps the user learn how to use this tool. In this prototype there is only one task agent (the drawing agent), but the final objective is to have several task agents, one for each GIS tool.

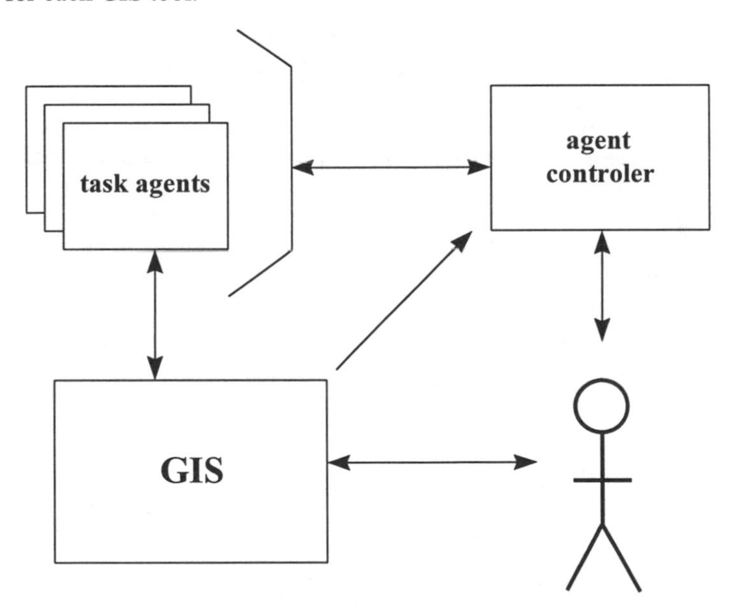

Figure 10.1 The agent architecture.

The agent controller monitors every event that occurs in the GIS and channels the relevant information to the appropriate task agent (currently, the appropriate agent is the only one that has been developed: the drawing agent).

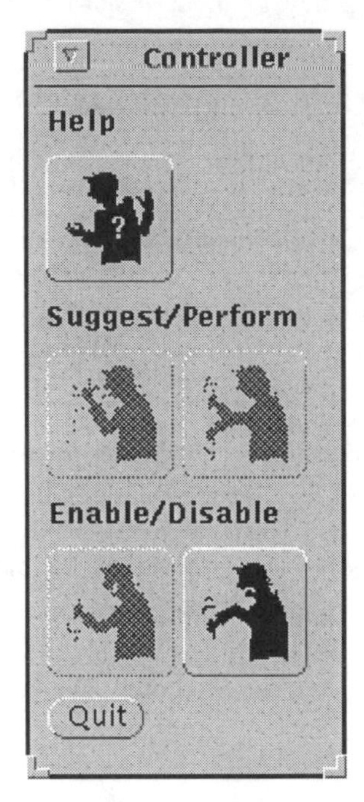

Figure 10.2 The agent controller user interface.

The task (drawing) agent receives event information from the controller and uses that information to review its state and re-assess the possibilities ahead. Afterwards, it will communicate to the controller its availability either to suggest or perform further actions. The agent controller includes a menu with six possible functions. These functions are presented to the user through icons that may be enabled or disabled, depending on the last communication with the drawing agent. The controller functions are as follows:

- *Help*. The Smallworld GIS manuals have all been converted into HTML form. Therefore it is possible, at any moment, to access the specific HTML page related to the part of the drawing tool that is being used. This icon is always enabled during the life of the controller.
- *Suggest*. The enabling of this icon by the controller means that the drawing agent knows what the state of the work is, and that it can suggest further actions to the user. If the user decides to click on this icon, directions will be provided.
- *Perform*. The enabling of this icon means not only that the drawing agents knows the state of the work, but also that it holds all the information to carry the execution of the function through to the end. If the icon is used, the current drawing function will be executed using the parameters that have already been entered.

- *Enable*. This function enables the controller to listen to events that are occurring in the GIS.
- *Disable*. This disables the controller. After the use of this icon the controller is 'asleep' and will not communicate either with the GIS or with the drawing agent. The Enable icon will 'awake' it and put it to work again;
- *Quit*. This kills the controller.

10.5.2 Further developments

This prototype architecture is hardwired to the GIS and the agent has been specifically created for the Smallworld drawing tool. There is no learning mechanism and no intelligence is attached to the system. It works quite well because this is a part of the system that, although very repetitive, needs very little personalisation. We would like to create more abstract task agents that would learn from the user how to automate different tasks (preferably spatial tasks). The aim is to create spatial task agents that can reason over spatial processes or data. This type of agent will need knowledge-based structures, including spatial knowledge as well as machine learning mechanisms.

ACKNOWLEDGEMENTS

This research is funded by PRAXIS XXI and JNICT through the fellowship PRAXIS XXI/BD/2920/94.

The authors would like to thank: Miguel Capitão of OBLOG Software, as one of the authors of the initial version of this chapter; Winton Davies of the Department of Computer Science, University of Aberdeen, Scotland, for reading earlier drafts of this chapter and supplying helpful comments; and Gillian Kendrick and all the Corporate Development and Product Marketing staff at SmallworldWide, Cambridge, for their help during the development of the drawing tool agent prototype.

REFERENCES

ADLER, R. M. and COTTMAN, B. H., 1989, A development framework for distributed artificial intelligence, in *Proceedings of The Fifth Conference on Artificial Intelligence Applications*, IEEE.

ARPA KSE, 1995, *KIF Manual*, ARPA Knowledge Sharing Effort, 1995, http://www.cs.umbc.edu/kse/kif/.

BATTY, M. and XIE, Y., 1994, From cells to cities, *Environment and Planning B: Planning and Design*, **21**, 31–48.

CENGELOGLU, Y., KHAJENOORI, S. and LINTON, D., 1994, A framework for dynamic knowledge exchange among intelligent agents, in *American Association for Artificial Intelligence (AAAI) Fall Symposium*, November.

CHESS, D., GROSOF, B., HARRISON, C., LEVINE, D., PARRIS, C. and

TSUDIK, G., 1995, Itinerant agents for mobile computing, in *IEEE Personal Communications*, **2**(5), October, pp. 34–49.

Communications of the ACM, 1995, *Special Issue on Digital Libraries*.

CYPHER, A., 1993, *Watch What I Do: Programming by Demonstration*, Cambridge, MA: The MIT Press.

FERRAND, N., 1995, Multi-reactive-agents paradigm for spatial modelling, *Proceedings of the GISDATA Workshop on Spatial Modelling*, Stockholm, June.

FININ, T., FRITZSON, R., MCKAY, D. and MCENTIRE, R., 1994, KQML, a language and protocol for knowledge and information exchange, in FUCHI, K. and YOKOI, T. (Eds.), *Knowledge Building and Knowledge Sharing*, Ohmsha and IOS Press.

FININ, T., LABROU, Y. and MAYFIELD, J., 1997, KQML as an agent communication language, in BRADSHAW, J. (Ed.), *Software Agents*, Cambridge, MA: The MIT Press.

FRANKLIN, S. and GRAESSER, A., 1996, Is it an agent, or just a program? A taxonomy for autonomous agents, in *Proceedings of the Third International Workshop on Agent Theories, Architectures, and Languages*, Springer-Verlag. `http://www.msci.memphis.edu/~franklin/AgentProg.html`

GOSLING, J. and MCGILTON, H., 1995, The Java language environment: a white paper, http://java.sun.com/docs/white/langenv/.

GRAY, R., KOTZ, D., CYBENKO, G. and RUS, D., 1997, Agent Tcl, in COCKAYNE, W. and ZYDA, M. (Eds.), *Itinerant Agents: Explanations and Examples with {CD-ROM}*, Greenwich, CT: Manning Publishing

GRAY, R. S., 1995, Agent TCL: a transportable agent system, in *Proceedings of the CIKM Workshop on Intelligent Information Agents, Fourth International Conference on Information and Knowledge Management (CIKM'95)*, Baltimore, Maryland, December.

GRUBER, T. R., 1993, A translation approach to portable ontology specifications, *Technical Report KSL 92-71*, Knowledge Systems Laboratory, Computer Science Department, Stanford University, September 1992, revised April 1993.

HARRISON, C. G., CHESS, D. M. and KERSHENBAUM, A., 1995, Mobile agents: are they a good idea? Research report, IBM Research Division, T. J. Watson Research Center, Yorktown Heights, NY 10598.

HUHNS, M. N., (Ed.), 1987, *Distributed Artificial Intelligence*, Los Altos, CA: Morgan Kaufmann.

IEEE Computer, 1996, *Special Issue on the Digital Library Initiative*, May.

JENNINGS, N. R., 1995, Agent software, in *Proceedings of The Agent Software Seminar*, UNICOM Seminars, London, 25-26 April.

KNASTER, S., 1995, *Magic Cap Concepts, Development Release 1*, http://www.genmagic.com, 10 September.

KOZIEROK, R. and MAES, P., 1993, A learning interface agent for scheduling meetings, in *Proceedings of the ACM–SIGCHI International Workshop on Intelligent User Interfaces*, Florida, January.

LANGE, D. and CHANG, D., 1996, Programming mobile agents in Java: a white paper, September, http://www.trl.ibm.co.jp/aglets/whitepaper.htm.

MAES, P., 1994a, Agents that reduce work and information overload, *Communications of the ACM*, **37**(7), July.

MAES, P., 1994b, Modelling adaptive autonomous agents, *Artificial Life*, **1**(1/2), 135–62.

MINAR, N., BURKHART, R., LANGTON, C. and ASKENAZI, M., 1996, The Swarm simulation system: a toolkit for building multi-agent simulations, http://www.santafe.edu/projects/swarm/overview.ps.

MINSKY, M., 1986, *The Society of Mind*, New York: Touchstone, Simon & Schuster.

NASA, 1995, *CLIPS, a Tool for Building Expert Systems*, NASA Johnson Space Center, http://www.jsc.nasa.gov/~clips/CLIPS.html.

OUSTERHOUT, J. K., 1993, *Tcl and Tk Toolkit*, Reading, MA: Addison-Wesley.

QUASAR KNOWLEDGE SYSTEMS, 1995, *SmalltalkAgents*, http://www.qks.com.

RODRIGUES, A., RAPER, J. and CAPITÃO, M., 1995, Implementing intelligent agents for spatial information, in *Proceedings of The Joint European Conference on Geographical Information*, The Hague, The Netherlands, 26–31 March.

RODRIGUES, A., GRUEAU, C., RAPER, J. and NEVES, N., 1998, Environmental planning using spatial agents, in *Innovations in GIS 5*, London: Taylor & Francis.

TOOMEY, C. N., SIMOUDIS, E., JOHNSON, R. W. and MARK, W. S., 1994, Software agents for the dissemination of remote terrestrial sensing data, in *Proceedings of the Third International Symposium on Artificial Intelligence, Robotics and Automation for Space* (i-SAIRAS 94), NASA Jet Propulsion Laboratory, Pasadena, California, 18-20 April.

WAYNER, P., 1994, Agents away, *Byte*, May, 113–18.

WAYNER, P., 1995a, Agents of change, *Byte*, March, 95.

WAYNER, P., 1995b, Free agents, *Byte*, March, 105–14.

WAYNER, P., 1995c, Agents unleashed: a public domain look at agent technology, *AP Professional*, Boston, MA 02167.

WOLFRAM, S., 1994, *Two-dimensional Cellular Automata, Cellular Automata and Complexity: Collected Papers*, Reading, MA: Addison-Wesley.

WOOLDRIDGE, M., 1995, Conceptualising and developing agents, in *Proceedings of The Agent Software Seminar*, UNICOM Seminars, London, 25–26 April.

WOOLDRIDGE, M. and JENNINGS, N. R., 1995, Intelligent agents: theory and practice, *Knowledge Engineering Review*, **10**(2).

Virtual Reality

CHAPTER ELEVEN

Virtual reality, the new 3-D interface for geographical information systems

Menno-Jan Kraak, Gerda Smets and Predrag Sidjanin
Delft University of Technology,
PO Box 5030, 2600 GA Delft, the Netherlands

11.1 INTRODUCTION

Developments in the world of spatial data handling show a clear increase in the diversity and size of GIS databases. Those databases now also include large sets of three-dimensional geometry and related attribute data. When starting a GIS project, researchers, among others, are interested to learn more about the data potentially to be used. Questions such as 'What do we have?' and 'How does it look?' are frequent. GIS vendors offer data viewers to browse and query the database, such as ESRI's ArcView and Intergraph's Vistamap. However, these cannot handle three-dimensional data. The emerging Virtual Reality (VR) technology does operate in a three-dimensional virtual world. It offers the viewer a set of stimuli of the 'real three-dimensional world', and provides a high level of spatial cognition, when interactivity is provided to freely navigate the virtual world, and this world is visualized sophisticatedly enough to perceive a 'real' environment (Edwards, 1992). The research question challenged here was 'What has Virtual Reality to offer the world of spatial data handling when it come to the exploration of three-dimensional spatial data sets?' The approach followed was to look at VR as an interface to a subset of the GIS database, and to implement GIS functionality on the level of a data viewer. This chapter reports on this experiment.

From the literature, several examples of links between GIS and VR are known: see, for example, McGreevy (1993), Raper *et al.* (1993), Camara and Neves (1995), Dias *et al.* (1995), Neves *et al.* (1995) and Schee and Jense (1995). They show that it is possible to create a virtual GIS world, the user being immersed in this world, in which he or she has to navigate and initiate spatial operations. Some important questions remain: 'Can we handle it?'; 'How will all this information be perceived?'; 'What will be the possible impact of this interface on the "map" in its function to explore, analyze and present spatial data?'

To test some of the ideas on the interaction of VR and GIS, and to answer the questions raised above, a virtual GIS world of the Delft University of Technology's campus area has been created. It consists of the area's basic topography, including a three-dimensional component. Generic data on the University's Faculties are used as attribute data. The functionality of the virtual GIS world interface can be compared with GIS data viewers. These offer options to browse and visualize spatial data sets. More advanced GIS functionality will only be added when it has been proved that a VR interface to the GIS database improves the understanding of spatial relations.

The current research project is a collaborative effort of the Faculties of Architecture, Industrial Design, Technical Mathematics and Informatics and Geodesy of the Delft University of Technology.

11.2 THE DATAVIEWER CONCEPT AND VIRTUAL REALITY

If one wants to understand spatial relations, geographical information systems cannot do without visualization tools. Visualization is employed to explore, to analyze and to present spatial data. Among the tools offered by vendors to allow for those different visualization approaches are the data viewers. Data viewers lack, purposely, the huge amount of (analytical) GIS functions available to the full GIS packages. They offer function to browse, query and present spatial data sets. Often, they are equipped with a relatively simple Graphical User Interface (GUI).

Nevertheless, many GIS users have difficulties in accessing their spatial data. Today, the most sophisticated GUI interfaces offered by GIS vendors are of the Windows, Icons, Mouse and Pop-up Menus (WIMP) type. However, current human-computer interface research is working on alternatives to this approach (Brown, 1994; Davenport, 1994). Virtual Reality (VR) offers an even more natural interaction with the data, and promises to be an interesting alternative here. Since VR is about the production of a simulated physical world, and some GIS databases contain three-dimensional data about this world, a combination of both technologies is obvious. When the GIS database stores the spatial data's geometric component in three-dimensional co-ordinates, a virtual model of the study area can be created, and the user can be given the opportunity to become fully immersed into the database. Navigation options still allow for a traditional map (over)view. The spatial data's attribute component can be used to 'dress' the virtual model, e.g. these data can be used to give the model its colour. The result will provide the GIS users with a completely new view on their spatial data.

The effectiveness of such an interface depends on several factors. These include the type and experience of the user, the type of problems to be solved, the need for interaction, and the possibility of using different kinds of 'interface strategies'. In the experiment discussed here, the users worked with a familiar environment, their own university campus area, and they used the system just to browse the data available. Interaction is inherent to VR and the different navigation modes allow for different approaches to the data.

To avoid the WIMP interface approach, with screens full of icons and drop-down menus, a limited number of virtual tools have been designed. The look of the tools, as well as their use, should initiate intuitive user action. The next section will describe them in more detail.

11.3 BUILDING THE PROTOTYPE

The virtual reality system employed during the research was an IBM Project Elysium model Immersive Virtual Reality 4. The key components of the system include the V-Flexor hand-held control device, a Visette 2 head-mounted display and a DOF tracker (see Figure 11.1). This last device translates the whereabouts and movement of the user to the system. The head-mounted display offers the user a stereo view of the virtual model. It is also capable of providing stereo sound. Movement of the user's head will immediately affect the user's view of the model. For instance, turning the head to the left will also rotate the displayed model to the left. The hand-held control device allows the user to control the speed of movement. With the virtual gloves seen on the display, which is directly linked to the V-flexor, the user can touch objects and activate (GIS) functions. Together with the software provided with the system, the package V-Space was used to build the virtual model out of a GIS data set. V-Space is an interactive three-dimensional modelling system and provides functionality to create fully rendered texture-mapped images and animations. The data structure applied includes a hierarchical display which makes it possible to display a single object at different levels of detail, depending of the distance to the viewer. The system software includes a graphics and sound library, and a tracker and control library to interact with the head-mounted device and the hand-held control device.

The geometry of the VR model of the Delft University campus area has been converted from an Arc/Info coverage. The coverage contained data collected by photogrammetric techniques from aerial photographs. Among the data, the outline of each building was used. Most buildings were split into several parts, between which a significant height difference could be observed. Heights were originally stored in the corresponding attribute tables. Next to this coverage, a data set with generic topography, with features such as roads and canals, was used. The data were exported from the Arc/Info database, and via a reformatting programme imported in the VR model, which is displayed in Figure 11.2a. The resulting buildings all had primitive 'cube-like' shapes. Some of the characteristic buildings in the campus area would not be recognised when composed of these simple geometric shapes, and were modelled individually using the VR software. During the conversion process the building identifiers were kept and could be linked to a colour table, which allows one to colour the VR model on the basis of attribute data. These identifiers are also the key link to the attribute data that is currently included in the system.

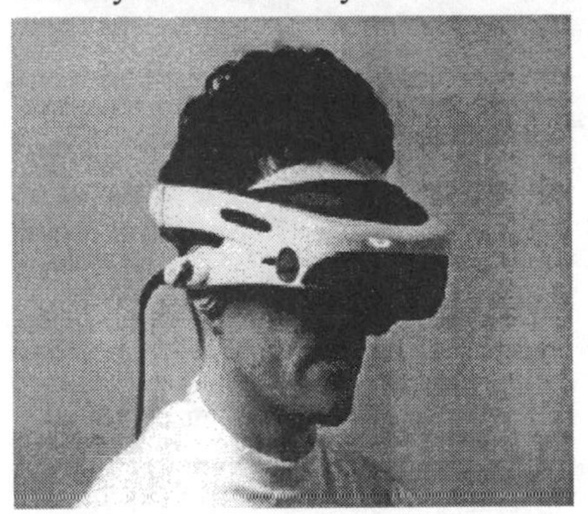

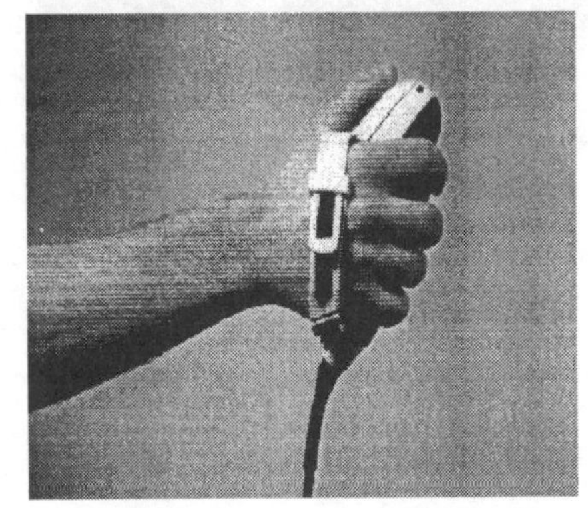

Figure 11.1 The components of the VR hardware used: (a) the head-mounted display and (b) the hand-held control device.

11.4 GIS FUNCTIONALITY IN THE VR MODEL

Those GIS functions included in the VR system are represented by iconic three-dimensional tools. The tools are organized on a plane which is shown in Figure 11.2b. The plane can be reached by the user by bending his or her head downwards, and the individual tools can be grabbed from the plane by the hand-held control device, represented by a glove in the virtual model. With the tools, the user can touch objects in the virtual model and activate their function. The plane is available in any navigation mode and at any location. By sound feedback, the user is informed about the success of a particular action or function.

In the virtual GIS world, two navigation modes exist. It is possible to fly or walk through the area. Flying allows one to get either a bird-eye view or the traditional 2-D map view of the campus area. Walking is represented by the tool *clogs* and flying by the *balloon.* To test the interface potential, some generic GIS functionality has been incorporated. The user can measure distances, areas and volumes. These operations are based on the model's geometry. It is also possible to query the model or parts of the model. One can touch a building and retrieve basic statistics, such as the number of staff and students. This type of query is hypertext based, and will, in the future, activate predefined links to the GIS database, or multimedia components such as sound and images. Currently, only a small subset of the GIS database is used.

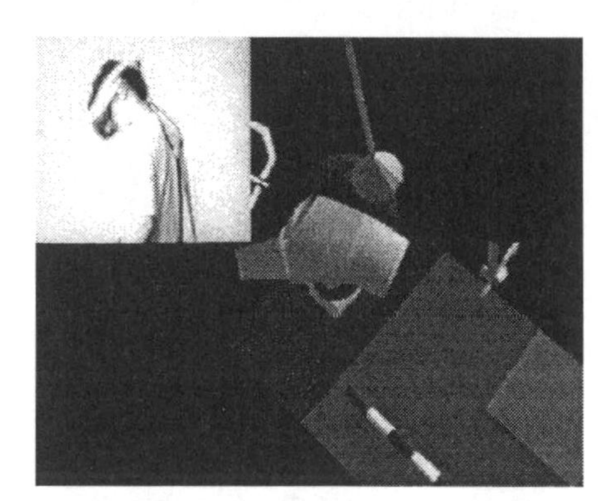

Figure 11.2 The Delft University campus area (a), and the functional tools (b).

To define distances, three tools are available to the user. A *stick* can be used to measure the shortest distance between two objects in the model (Figure 11.3a). A *stepmeter* is available to measure the shortest distance between two objects along the road network incorporated in the model. A *pendulum* can be used to measure heights of individual buildings (Figure 11.4a). Area measurements are limited to determining the size of building floors, and the function is initiated by activating the *square* tool (Figure 11.3b). Volumes can be defined by use of the *cube*, the tool that calculates the volume of objects touched (Figure 11.4b).

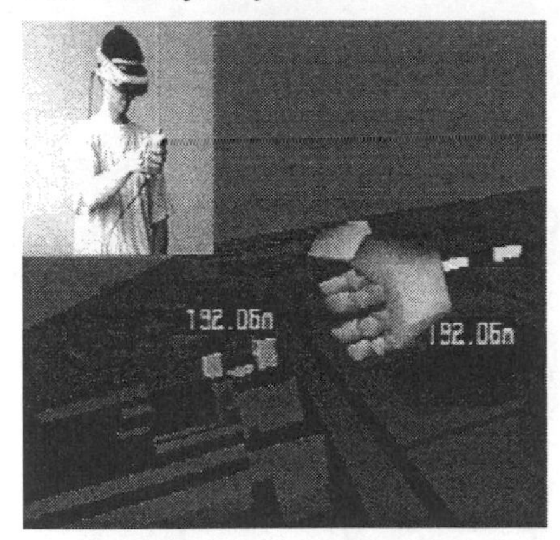

Figure 11.3 The measurement of distances (a) and areas (b). Both illustrations show the virtual representation of the hand-held device.

Working with attribute data is not yet implemented. The basic idea is to realize this option by virtual hypertext. A tool in the shape of the white letter *i* on a blue square (information sign) would activate the system's hypertext mode. When the user touched an object, its identifier, such as, for instance, its type or name, would appear. Touching this text would open a spreadsheet which, in the case of a building, would be displayed on one of its walls. The spreadsheet would show the overview of data known to the system and the green hypertext words would lead the user deeper into the data, and allow further exploration. In the case of the campus information system, generic data on buildings, faculties, staff and students would be available.

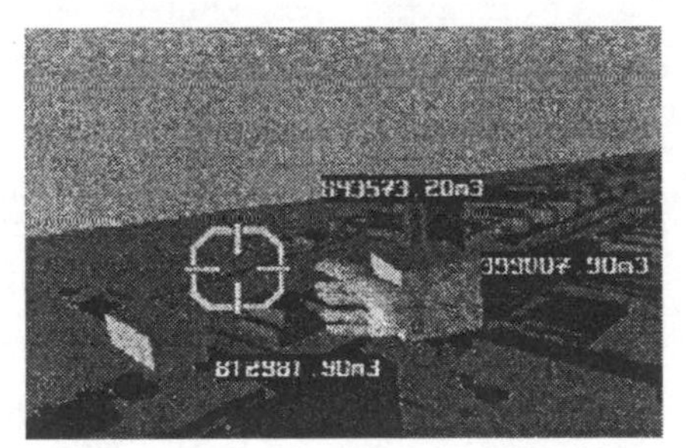

Figure 11.4 The measurement of heights (a) and volumes (b).

The next research phase will also be devoted to the integration of textured photographs of the objects in the models. When navigating through the model, the user will experience a more realistic environment, which will be especially useful in walking mode. This approach coincides with another ongoing research project in the field of close-range photogrammetry, that uses images obtained via a digital camera to produce a correct geometrical fit on CAD models of the buildings photographed. Working with the current hard- and software, the inclusion of images will increase the amount of data to be processed and probably slow down the system's interactivity. However, here the software's hierarchical object storage capabilities will be applied.

11.5 CONCLUSIONS AND FUTURE RESEARCH

The research reported here experimented with VR as an interface to GIS. The generic concept was to observe the combined VR/GIS functionality on the level of GIS data viewers. However, the VR approach offers the user an interactive three-dimensional view on the data. Most users viewing the system were sceptical at first when they saw the primitive building blocks, but as soon as they were immersed in the model they were surprised by its effect. Currently, only geometric functions are active, but the subset extracted from the GIS database already includes attribute data, and one of the future tasks will be to extend the system functionality in this field. A video is available to demonstrate the model in action. Future research will concentrate on the effectiveness of the system with regard to user tasks.

REFERENCES

BROWN, P., 1994, The ethics and aesthetics of the image interface, *Computer Graphics*, **28**(1), 328–31.

CÂMARA, A. and NEVES, N., 1995, Exploring virtual exosystem. Workshop material 'GIS and multimedia to explore reality', *JEC on GIS,* Den Haag.

DAVENPORT, G., 1994, Bridging across content and tools, *Computer Graphics*, **28**(1), 31–3.

DIAS, A.E., *et al.*, 1995, Virtual environments logging using BITS, in *Proceedings of the 1st Conference on Spatial Multimedia and Virtual Reality*, Lisbon, pp. 61–72.

EDWARDS, T. M., 1992, Virtual world technology as a means for human interaction with spatial problems, in *Proceedings of GIS/LIS'92*, San Jose, ACSM-URISA-AM/FM.

MCGREEVY, M. W., 1993, Virtual reality and planetary exploration, in WEXELBLAT, A. (Ed.), *Virtual Reality—Applications and Explorations*, Cambridge, MA: Academic Press.

NEVES, N., *et al.*, 1995, Virtual GIS room, in *Proceedings of the 1st Conference on Spatial Multimedia and Virtual Reality*, Lisbon, pp. 45–53.

RAPER, J., MCCARTY, T. and LIVINGSTONE, D., 1993, Interfacing GISwith virtual reality technology, in *Proceedings of the AGI Conference*, Birmingham: AGI.

SCHEE, L. H. VAN DER and JENSE, G. J., 1995, Interacting with geographical information in a virtual environment, in *Proceedings of JEC-GIS*, The Hague, Vol. 1, pp. 151–7.

CHAPTER TWELVE

A real-time, level-of-detail editable representation for phototextured terrains with cartographic coherence

Joaquim Muchaxo, Jorge Nelson Neves and António S. Câmara
Virtual Reality Lab, Environmental Systems Analysis Group, New University of Lisbon, 2825 Monte de Caparica, Portugal

12.1 INTRODUCTION

The techniques presented herein aim to represent data defined on R2. For techniques on rendering terrain data defined on the sphere, the reader is referred to Schroder *et al.* (1995).

Extensive research has been carried out on triangulation techniques for height fields, mainly for flight simulators (Hughes, 1993) and geographical information systems (GIS) (Ware and Jones, 1997). Previous work was aimed at creating a TIN that approximates the terrain either by decimation of data points (Schroeder *et al.*, 1992) or by hierarchical triangulation (DeFloriani and Puppo, 1988). Implementations in the last category provide local control of the level of detail. However, these triangulations are computationally intensive, do not allow fast spatial indexing and are hard to edit. Regular triangulations have also been presented (Hughes, 1993). They are simple, but require many more triangles than a TIN for the same level of accuracy.

Some of the characteristics that are desirable for the continuous level of detail rendering of phototextured terrains include the following:

- the generation of images according to pre-established error bounds for geometry and texture detail (Lindstrom *et al.*, 1996)
- the creation of a triangulation with cartographic coherence (i.e. not generating edges that contradict terrain features) to ensure a simplification with an almost optimal number of triangles (Scarlatos and Pavlidis, 1992)
- the reduction of texture data to its optimal value: the size of the screen area occupied by the projection of the terrain
- the smooth junction of meshes with different resolutions
- the storage and management of the phototextured terrain in multi-resolution form and according to spatial influence, allowing fast retrieval or editing at an arbitrary

resolution

More recently, triangulations based on the quad tree (Samet, 1984) partitioning of the terrain data were proposed. Gross (1995) proposed the usage of a wavelet representation to triangulate terrains. This uses detail signals obtained from the inverse wavelet transform to decide whether or not to include a particular vertex in the triangulation. However, this technique does not check the generation of images according to pre-established error bounds. Implementations have been presented that do check this property, but that do not use wavelets (Muchaxo, 1995; Lindstrom *et al.*, 1996).

To avoid cracks in the surface, many implementations build triangulations with subdivision connectivity (Gross *et al.*, 1995; Schroder *et al.*, 1995). These triangulations add many edges that contradict the critical lines formed by terrain features. Others disable simplification on some locations using dependency information (Lindstrom *et al.*, 1996), thus not reducing the number of polygons efficiently.

12.2 IMPLEMENTATION ISSUES

The method starts with a data matrix, which is later converted to multi-resolution form using wavelets defined on the Haar basis. The Haar basis was chosen because it is fast to compute, and it is simple to organize detail coefficients according to their spatial influence. Berman (1994) describe the usage of an *N*-step wavelet transform organized in a quad tree for multi-resolution painting of images. The reader is referred to Gross (1995) for an explanation of where to find the various frequency channels (i.e. the coefficients for a particular level) inside the transformed matrix.

The outline of the quad tree presented here is similar to that of Berman (1994), except that the quad tree is not expanded until the data point. In 3-D rendering systems, the rendering is done on a polygon and texture basis. Hence the levels of detail are stored and managed on a quad cell basis using small submatrices. The sizes of these ($N*M$) are chosen on the basis of the system used: the optimal texture block size for the graphics engine and the optimal number of polygons of a geometric entity (around $N*M/4$ in this implementation).

Some data types (layers) have several independent components such as textures (RGB): this implementation uses the same quad tree to manage them. To save memory for simulation applications, one might not read all the tree from disk. The pointers to the children might be file offsets instead of pointers to memory locations (which require four additional flags). To save more memory, one can use sparse matrices for the coefficients, so that the coefficients that are zero are omitted.

For a better understanding of the representation used the data struture proposed for the quad tree is presented next:

```
type layer = record
    Img: array [ncomponents][NxM] of float
    Root: pointer to Wavelet_Node
end record
type Wavelet_Node = record
    Coefs: array [ncomponents][3] of pointer to coefficients_array
    Children: array [4] of pointer to Wavelet_Node
    parent: pointer to Wavelet_Node  // null if parent = root
```

```
    min_value, max_value: float
    leaf_node: pointer to lod_info
end record
type coefficients_array = array [ncomponents][N*M] of float
type lod_info = record
    Img: array [N][M] of float
    m: array [N][M] of float
    updated: boolean
    previous, next: pointer to lod_info
end record
```

The leaf_node field is not present in the quad tree saved on disk. It is used to store level-of-detail information in real time. It is explained in the next section, along with the min_value and max_value fields.

12.3 ESTIMATION OF PERSPECTIVE CULLING

To reduce real-time computation, two projection factors (P_{FG} for geometry and P_{FT} for textures) are estimated that indicate the reduction of detail involved on the perspective projection. Lindstrom *et al.* (1996) considered that the elevation error resides along the vertical axis, regardless of the orientation of the vertex normals.

As Lindstrom *et al.* (1995) describe, this technique always considers, for LOD calculations, the plane normal to the vector that connects the viewpoint and the centre of the quad cell instead of the view plane.

The culling that depends on distance can be evaluated simply and accurately and is equal to C_d:

$$C_d = \frac{d_f}{\min(d_{TIN \to VP} - r_{TIN}, \rho)} \qquad (12.1)$$

where d_f is the focal length, $d_{TIN \to VP}$ is the distance from the TIN to the viewpoint, r_{TIN} is the radius of the TIN and ρ is the near side clipping value.

The projection factors are functions defined on multiple branches. These branches correspond to the areas defined in Figure 12.1. First, some variables must be defined. cos(θ) is given by

$$\cos(\theta) = v * n_{avg} \qquad (12.2)$$

where v is a unit vector that points from the viewpoint to the quad cell's centre and n_{avg} is the average of the polygon normals of the mesh; $\cos(\alpha)$ is the smaller dot product between n_{avg} and the triangle normals. It is used to define a cone, the peak of which is the centre of the quad cell and that encloses all of the polygon normals. The projection factors are given in Table 12.1.

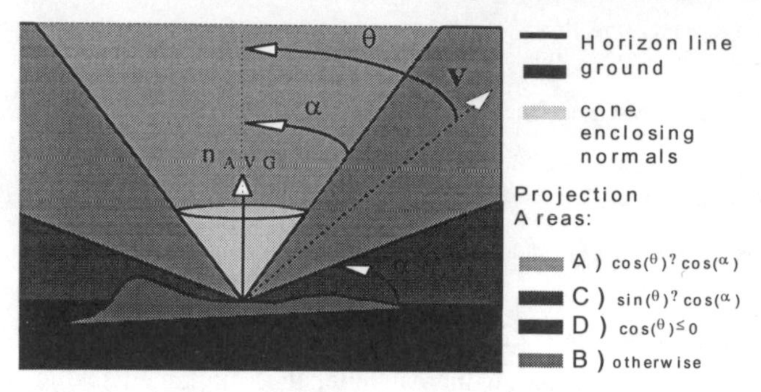

Figure 12.1 A cell of the terrain and the different areas used to determine the projection factors.

Table 12.1 Corrected projection factors.

Area	Report	Value of C_G	Value of C_T
A	v is close to n_{avg}	$\sin(2\alpha)$	1
B	$v.\ n_{avg} \cong 0$	1	$\sin(2\alpha)$
C	The TIN is hidden	0	0
D	Pf must be given a value that ensures continuity	$\sin(\theta+\alpha)$	$\cos(\theta+\alpha)$

For area A, the quad cell is being seen from above. Thus, the texture is seen at its full extent ($C_T = 1$) and the geometry reduction is small considering that the normals make, at most, an angle equal to α with the view direction. The inverse situation happens in area B. From area C the terrain is not visible, so the projection factors are equal to zero. In area D, functions were selected to ensure continuity.

The actual projection factors (P_{FT} and P_{FG}) used for a quad cell are given by

$$P_{FT} = C_d C_T, \quad P_{FG} = C_d C_G. \qquad (12.3, 12.4)$$

12.4 MANAGEMENT OF LEVEL OF DETAIL

The images are rendered according to two pre-established parameters: texture detail and geometry detail. The former is the number of pixels that correspond to a texel. The second is the projected elevation error, also measured in number of pixels.

In real time, each branch of the wavelet quad tree is expanded independently to meet the optimal level of detail for an area. To keep track of this expansion, the *leaf_node* field is non-null if the node is a leaf (there is exactly one leaf node on the path from a root to a terminal node). The *leaf_node* contains the texture (or elevation grid) for a quad cell, and pointers to the left and right leaf nodes for fast traversal of all leaf nodes.

Linear operations can be performed on either the topography or the texture (Berman *et al.*, 1994). If a linear operation is performed on any of the pixels (which is indicated by the flag *updated*), the coefficients of the nodes hierarchically below should be updated (Berman *et al.*, 1994). Instead, the linear operations ($y(i,j) = m(i,j).x(i,j) + b(i,j)$) are composed, i.e. $m(i,j) = m'(i,j)*m''(i,j)$, and the update is delayed until the resolution of the image must be increased or decreased. Note that it is not required to add the b value (Berman *et al.*, 1994). If the resolution must be increased, then only the coefficients of the current node are multiplied by m and the cache is scaled twice, broken into four and assigned to the children of the node. If the resolution must be decreased, then the coefficients of all nodes hierarchically below are multiplied by m (if $m<>1$). This is the only fundamental difference between this implementation and that of Berman *et al.* (1994) regarding the wavelet quad tree.

The values *min_value* and *max_value* are used to determine whether the quad cell data matrix at a particular level of detail has appropriate resolution for the current viewpoint. These values were pre-computed when creating the wavelet quad tree and are updated when some data value is edited. They are the minimum and maximum values of the data matrix for the quad cell. The difference between these values is projected with P_{FG}. The projected value is then verified to be smaller than the pre-established error bound $\tau = 4*G_d$, because a TIN will be used to eliminate the remaining error. Thus the usage of TINs generally reduces the depth at which a suitable quad cell can be found.

12.4.1 Construction of TINs according to terrain features

A TIN is created for each leaf quad cell. The TIN is constructed starting with two triangles forming the quad cell. Subsequently, each triangle is scanned using a special scan line algorithm that finds the enclosed data point or edge point that deviates more from the surface of the terrain. This point will be used to break the triangle in two or three triangles. Recursive subdivisions are made until the error involved in the triangulation is less than the pre-established error bound. The triangles are kept on a hierarchical tree to speed up point membership tests.

A division point over an edge is preferable to a point inside, so that the terrain's topological features are captured. Frequently, the edges do not cross any data point; to compensate for this effect, additional scan lines are processed in between, taking only edge points into consideration. Also, points over small edges must be ignored to avoid slivery triangles. Thus, for deviation comparison tests, the deviation value for an edge point is multiplied by a factor λ related to the edge size:

$$\lambda = \frac{(E - E_S)}{(E_L - E_S)} \cdot \frac{E_L}{E_S} \cdot C + 1 \qquad (12.5)$$

where E is the size of the edge on which the point was found, and E_S and E_L are, respectively, the sizes of the largest and the smallest edge of the triangle. C is a coefficient with a value of 0.5. Very small edges are not broken to avoid running out of floating point precision.

When the point chosen to break a triangle is over an edge, the triangle that shares that edge must also be divided by this point, thus avoiding gaps. Each of these points is given a unique code and stored in a hash table. This code is built up from the vertices that the edge connects. The same point might be chosen to break the confining triangle, but no other, since it must be the point that deviates more from the terrain. When that happens, the point is removed from the hash table. After all triangles are processed, those with edges on the hash table are broken into two. This is a simple scheme to avoid the management of complex data structures that keep track of neighbours, as happens in many TIN implementations (DeFloriani and Puppo, 1988; Scarlatos and Pavlidis, 1992). This representation guarantees cartographic coherence without having to consider several triangle split strategies, as described by Scarlatos and Pavlidis (1992).

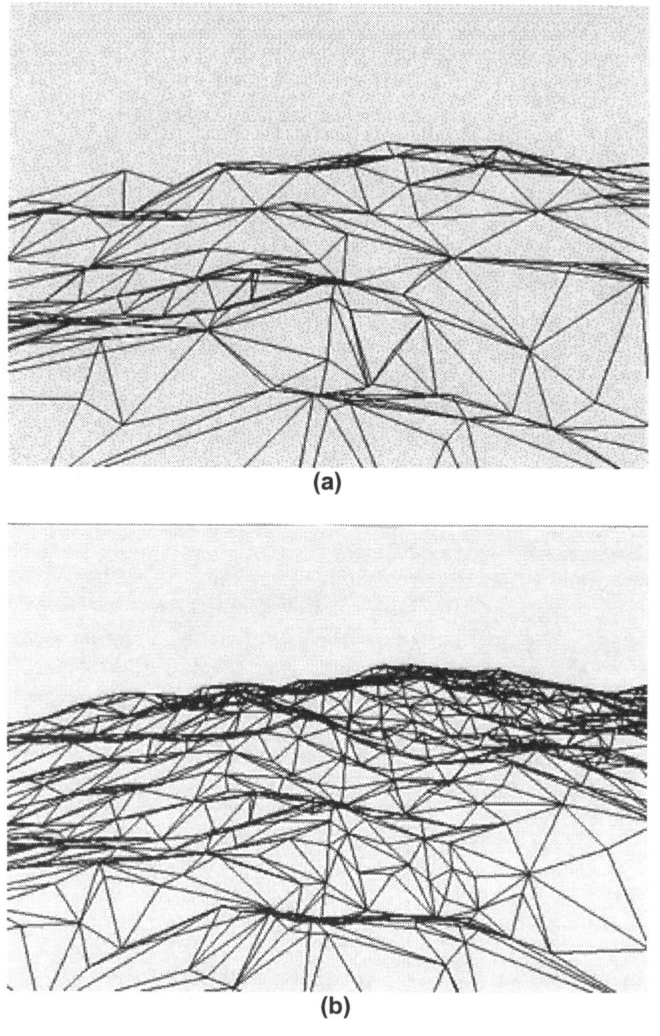

Figure 12.2 **TINs with a high degree of cartographic coherence: (a) $G_d = 4$; (b) $G_d = 1$.**

12.4.2 LOD rendering of textures

To reduce the amount of texture data to the size of the screen area occupied by the projection of the terrain, it is not sufficient to consider the normal culling relative to distance. One must also take advantage of the fact that textures seen from oblique angles occupy smaller areas than those seen upright.

To make better use of the limited texture memory available, several textures can be placed on the same texture block using mapping. The spaces inside texture blocks are managed using a first fit strategy. Also, textures are transposed if they fit on to a texture block that way, performing additional transformations on mapping co-ordinates.

To calculate the reduction for the second axis, the projected extents of the bounding box of the quad cell are computed. Then, on the texture axis that has the smaller projection, the number of texels used for that dimension is given by $\min(E_O + 2E_H\ T)$, where E_O is the projection of the corresponding extent and E_H is the projection of the extent at which the elevation of the terrain lies.

12.4.3 Joining cells with different resolutions

The technique designed is general in the sense that it can be used to ensure continuity of surfaces composed of several meshes of arbitrary topological type. For a vertex on a quad cell boundary, four directions (*NW*, *NE*, *SW* and *SE*) are defined, where a quad cell must exist that shares the vertex. As an example, a vertex on the right boundary of a quad cell has both flags *NE* and *SE* reporting to that quad cell.

To rapidly detect vertices that are not shared, a unique key is given to each vertex as a function of its location and the vertices are stored on a hash table along with their flags. Whenever a mesh is created, modified or deleted, the flags of the boundary vertices involved are updated. Since the triangles of a mesh share vertices, one must not process the vertices more than once. To ensure this, a small temporary hash table is built with the vertices already processed.

If the flags of a vertex are all false, the vertex will cease to exist in the terrain geometry; hence it is removed from the hash table. If they are all true, then the vertex is shared by all confining quad cells; hence it is removed from the hash table since it needs no special care.

Before a rendering step, only the non-shared vertices are on the hash table. For each vertex, the quad cells that do not share it are identified by the vertex itself and by its flags. They can be determined very rapidly due to the hierarchical spatial subdivision of the quad tree. For each quad cell, the triangle where the vertex lies is also determined quickly using the hierarchical TIN and then broken into two to accommodate that vertex.

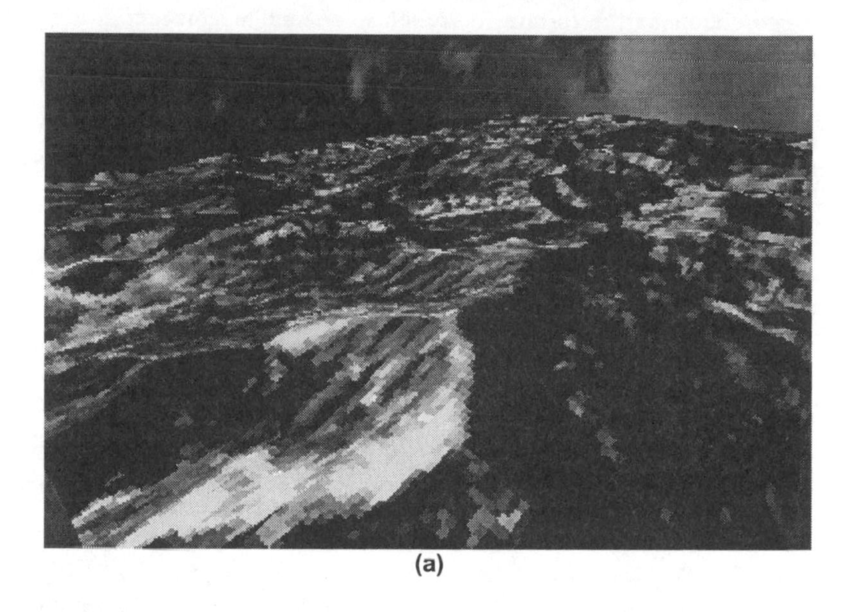

(a)

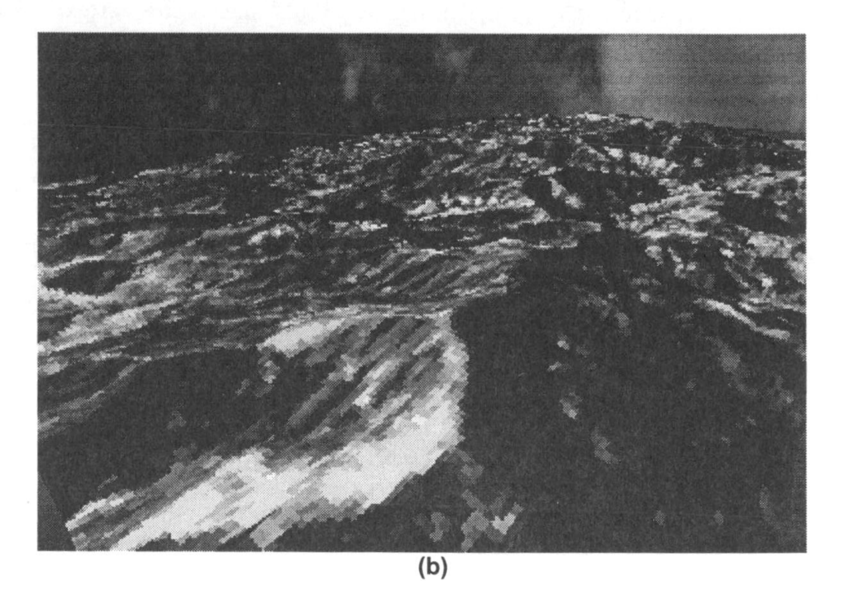

(b)

Figure 12.3 **Images rendered for different values of geometry detail (G_d) and texture detail (T_d): (a) $G_d = 1$, $T_d = 2$; (b) $G_d = 1$, $T_d = 1$.**

12.5 RESULTS AND CONCLUSION

This research was implemented on a 60 MHz Pentium™ computer with an SPEA FIRE™ accelerator graphics board. The software used was WorldToolkit™ 2.0 from Sense8. The code is currently being ported to WorldToolkit™ release 6, running on top of Open-GL.

The LOD management was tested using an elevation matrix of 512 × 256 points and a texture of the same size with a resolution of 16 bits per pixel. The screen size was 384 × 256. The texture memory required never exceeded five times the screen area (for a field of view of 60°).

	-	-	Figure 3a	Figure 3b
Geometry detail, G_d	4	4	1	1
Texture detail, T_d	2	1	2	1
Quad cells	661	1024	634	1024
Polygons	1155	1545	1870	2830
Geometry reduction	0.0044	0.0059	0.0071	0.0110
Texture memory used (×1000)	73.4	159	70.1	153
Texture memory wasted (×1000)	16.2	25.9	14.6	24.8
Frames per second (on Pentium)	5.2	3.7	4.3	2.9

Table 12.2: Summary of Results Obtained

The maximum output rate of the hardware used is very small, about 6000 textured polygons per second. Another main difference between the method here presented and that of Lindstrom (1996) is that all the terrain data around the user is considered, and not just that in the field of view (60°). This allows a rapid response when the user looks in a different direction. Thus our geometry reduction factor is (0.0071/5 per mil or a ratio of 704) compared to the reduction ratio accomplished by Lindstrom (which is 343 for a threshold corresponding to $G_d = 2$, since we consider the effect of aliasing on the rendering of the images). For a summary of results obtained see Table 12.2.

REFERENCES

BERMAN, D. F., BARTELL, J. T. and SALESIN, D. H., 1994, Multiresolution painting and compositing, in *Proceedings of SIGGRAPH '94*, pp. 85-90.

DEFLORIANI, L. and PUPPO, E., 1988, Constrained Delauney triangulation for multiresolution surface description, in *Proceedings of the 9th IEEE International Conference on Pattern Recognition*, November.

GROSS, M. H., GATTI, R. and STAADT, O., 1995, Fast multiresolution surfacc meshing, in *Proceedings of IEEE Visualization '95*, pp. 135-42.

HUGHES, P., 1993, Terrain renderer for Mars Navigator, Tech Note 8, in WOLFF, R. S. and YAEGER, L. (Eds.), *Visualization of Natural Phenomena*, Santa Clara, CA: TELOS, The Electronic Library of Science, pp. 261-6.

LINDSTROM, P., KOLLER, D., HODGES, L. F., RIBARSKY, W., FAUST, N. and TURNER, G., 1995, Level-of-detail management for real-time rendering of phototextured terrain, Georgia Institute of Technology; Army Research Laboratory, ftp://ftp.gvu.gatech.edu/pub/gvu/tech-reports/95-06.ps.z.

LINDSTROM, P., KOLLER, D., RIBARSKY, W., HODGES, L. F., FAUST, N. and TURNER, G., 1996, Real-time, continuous level of detail rendering of height fields, in *Proceedings of SIGGRAPH '96*, pp. 109-18.

MUCHAXO, J., 1995, Multiscale representation for large territories, in *Proceedings of the 1st Conference on Spatial Multimedia and Virtual Reality*, Lisbon, pp. 61-72.

SAMET, H., 1984, The quadtree and related hierarchical data structures, *ACM Computing Surveys*, **16**(2), 187-260.

SCARLATOS, L. and PAVLIDIS, T., 1992, Hierarchical triangulation using cartographic coherence, *CVGIP: Graphical Models and Image Processing*, **54**(2), 147-61.

SCHROEDER, W., ZARGE, J. and LORENSON, W., 1992, Decimation of triangle meshes, in *Proceedings of SIGGRAPH '92*, pp. 65-70.

SCHRÖDER, P. and SWELDENS, W., 1995, Spherical wavelets: efficiently representing functions on the sphere, in *Proceedings of SIGGRAPH '95*, pp. 161-72.

WARE, J. M. and JONES, C. B., 1997, A multiresolution data storage scheme for 3-D GIS, in KEMP, Z. (Ed.), *Innovations in GIS 4*, London: Taylor & Francis, pp. 9-24.

CHAPTER THIRTEEN

A virtual GIS room: interfacing spatial information in virtual environments

Jorge Nelson Neves, Pedro Gonçalves, Joaquim Muchaxo and João P. Silva
Environmental Systems Analysis Group, New University of Lisbon, 2825 Monte de Caparica, Portugal

13.1 INTRODUCTION

The combination of geographic information systems (GIS) and visualization remains insufficiently unexplored. Both fields are solid in theory and practice and have available systems in the market. As they seem to complement one another, they rarely interact or exchange information, and when they do, some information is lost in the process (Robertson and Abel, 1993).

Some applications demand the combination of GIS and visualization, providing a motive for greater interchange between these systems (Robertson and Abel, 1993). Environmental applications are in this category. Data of this kind are used for decision-making and so have to be easy to understand, with intrinsic complexity removed.

The communication of environmental data to the audience is critical. Other actions, such as interrogating databases or manipulating parameters (to specify scenarios), should also be accessible in an highly usable interactive system.

The proposed solution to this problem consists of the integration of both systems in a working environment. This environment should add user-friendliness, interactivity and immersion to the visualization process, promoting a better insight into the data. This solution provides a new kind of interaction with spatial information in general, and with GIS in particular, using transparently integrated virtual environments and GIS systems. That integration is attained with communication of objects between applications, using interoperable objects technology.

13.2 BACKGROUND

This research overlaps several scientific areas; namely, geographic information systems, virtual environments and interoperable objects. As a result, the present review refers to these three topics.

13.2.1 Geographical information systems

Goodchild (1987) divided GIS into four parts: input, storage, analysis and output. Considering the remarkable progress in recent years regarding input, storage and output, mainly due to the breakthroughs registered in the computer and remote sensing industry, the analysis ring was set far behind in this development. Several authors imply the need for the development of a toolbox for spatial analysis, consisting of the basic tools and building blocks for simulation in the GIS framework. Operational tools for spatial analysis are indispensable for sophisticated GIS, and much research effort should be devoted to their development (Goodchild, 1987; Openshaw, 1987; NCGIA, 1989; Fisher and Nijkamp, 1992).

There are two broad and opposing classes of models within spatial information. One is the class of field-based models, where the information acts as collection of spatial distributions that can be formalized as a mathematical function from a spatial framework. The best example is the idealized model of a grid covering the study surface with information regarding patterns of altitudes, land use or rainfall.

The other class is entity-based models, in which the information space is considered as being populated by geo-referenced discrete, identifiable entities (or objects). These model types result in the categorization of GIS packages in two distinct groups, called vector and raster systems, according to the mechanism of data encoding and storage. These two types have evolved, each with certain inherent strength and weaknesses. Several authors have discussed the characteristics, merits and shortcomings of these traditional GIS classes (see Burrough, 1986).

To supplant these two systems, one has to start to view geographic information in a object-oriented perspective. An object-oriented model transforms an information space by dividing it into several individual objects (Worboys, 1994). In Mattos *et al.* (1993), an object is referred to as having several general proprieties: identifiable, relevant (being of interest), describable (having distinct characteristics). The last item is provided by static properties that define the object's behavioral and structural characteristics.

The assignment of Object Oriented (OO) characteristics to a GIS must follow this multi-faceted entity (Worboys, 1994). Some approaches and methods to a full object-oriented GIS are discussed in Worboys (1994), and some systems already include this technology; namely, Intergraph's TIGRIS (Herring, 1991), SmallWorld GIS (Chance *et al.*, 1990; Newel 1992) and Geo ++ (van Oosterom and van den Bos, 1989). However, these systems are not yet in widespread use and are restricted to specialized markets.

13.2.2 Virtual environments

Virtual environments (VE), or virtual worlds, are those that result from the interaction between the cognitive level of the human being (usually designated as mental maps) and the visual and audible images produced by the computers. A virtual world is a space deliberately designed by man, representing real or abstract spaces, in which objects exist, these being governed by rules specified for those worlds that they inhabit (Jacobson, 1994).

Virtual worlds can be used to organize, represent and manipulate data with a multi-dimensional nature. They can be seen as plane images, as 2.5-D models in a conventional monitor, or in a truly three-dimensional space in an immersive environment. Virtual worlds can also include information with more than three dimensions, including the temporal dimension (Jacobson, 1994).

The use of virtual reality techniques allows a spatio-temporal representation with great fidelity, because it is possible for the decision-maker to interact directly with the elements that form the basis of his or her decision. This ability to feel, as opposed to qualify, the results of an action makes the understanding clearer and universal.

The use of visualization methods in the analysis of geo-referenced data is based essentially on static models that, being so, restrict the visual analysis capabilities (Dioten, 1995). The use of virtual reality, which gives the user the ability to change viewpoints and models dynamically, can overcome that limitation.

With the new low-cost virtual reality (VR) hardware/software platforms, its application has been extended to several fields of science, from biology to architecture. The application of this technology to spatial information is based upon the exploratory characteristics of VR.

GIS are usually defined as a common base for information systems and the several fields of knowledge that use spatial analysis techniques. These techniques can be used as data input, storage, query, transformation and representation tools that represent the real world (Schee, 1995).

Virtual environments facilitate human-computer interaction by the use of a three-dimensional representation and direct manipulation of virtual objects (Burdea and Coiffet, 1994). In traditional VR systems, the user is immersed in a 3-D world generated by a computer using a head-mounted display and position/orientation sensors.

Fairchild (1993) and McGreevy (1993) have reported significant work in the use of visualization for information management and the use of VR for planetary exploration, respectively, and constitute fundamental references on the association of VR and GIS. The Sequoia 2000 project (Stonebraker, 1994) constitutes a good working basis in the visualization aspects of this association.

13.2.3 Interoperable objects

As GIS and VE to interface with spatial data can overlap, a communication model between the two systems should be developed, with objects exchanged via network or locally, in the most transparent way possible. This solution avoids re-inventing the wheel, with old GIS functions being developed outside the GIS, and also the burden of adding new interfaces and visualization capabilities to closed systems, as GIS have always proven to be.

To develop a communication model, generic data types and operation types must be defined for each type of data interchanged. This is the communication interface.

To exchange data, the network binary format must also be agreed, because it is unlikely that applications will be compiled and linked in the same way. Using low-level communication protocols such as TCP (stream sockets, point-to-point connections) or UDP (data gram sockets), the binary format must be defined and managed by the user. If RPC (remote procedure calls) or higher-level protocols are used, only the communication interface needs to be defined.

There is a concern in the major software developing companies to create standards to allow applications to access components created by different software vendors. These standards involve a technology that is most often called 'interoperable objects' (Valdés, 1994a). The principal variants are Microsoft's Component Object Model (COM) (Williams and Kindel, 1994), which is the base of OLE 2.0 (Brockschmidt, 1994), and IBM's System Object Model (SOM) (Campagnoni, 1994) which are already implemented for various platforms.

A meeting place in these technologies is the use of an interface definition language (IDL). This allows a language-independent expression of interfaces. Using an IDL compiler, the interface is mapped for a specific language, allowing clients and servers to be implemented in different languages.

Interoperable objects IDLs generally impose no restriction on object types, because their origins are IDLs for RPC, a technique that enables the exchange of any kind of data structure over the network. They are object-oriented extensions of these languages and enable the creation of objects in such a way that an application can be a client and a server at the same time.

Another common characteristic of interoperable objects is that they provide a standard binary format. No source code is necessary to subclass objects, and use them to build applications, possibly in a language different from the one in which they were initially created. A server's implementation can be corrected or extra functionality added without recompiling client programs (Valdés, 1994b).

13.3 PREVIOUS WORK

The use of virtual environments to visualize surface and subsurface spatial data, in more abstract terms, 2.5-D or 3-D data, has gained widespread attention from the scientific community in recent years. The work from the Georgia Institute of Technology is worth mentioning, specifically the VGIS project. This system is not so general as the one proposed here, because it is application-oriented, it is an application to the military field, and so it was not designed to incorporate general spatial information data types with which to interface.

Schee and Jense (1995) described a system that could be considered similar to ours in the fact that they also intended to make a bridge between VE and GIS. They considered that the visualization and manipulation of spatial data using VE poses some requisites in the interface between both systems, determined by factors such as:

- the VE system performance, i.e. the frame rate and the maximum amount of information that can be loaded
- the differences in information representation between the GIS and VE systems
- communication between both systems, maximum load and speed

Those limiting factors originate key decisions in the system design that we also considered it important to follow:

- the VE system cannot be limited by the processing capabilities of the GIS system or by the speed/rate of transmission, the VE system should continue execution even before the arrival of an answer to a request to the GIS system
- information coming from the GIS that is not relevant should be eliminated or not sent at all
- conversion between both systems should be included

Other key aspects of the VE application were also influenced by the work of other researchers. In navigation and locomotion in a virtual world, the work done at the University of Virginia was a strong influence, mainly with the introduction of the Worlds In Miniature (WIM) concept (Pausch and Burnette, 1995; Stoakley *et al.*, 1995). The CAVE environment (Cruz-Neira *et al.*, 1993) and the Nanomanipulator project (Taylor *et al.*, 1993) were particularly important for information presentation and other general interface issues.

13.4 SYSTEM DESIGN

Our experience with virtual environments showed that some daily tasks are not easily replicated in a virtual world, with poor performance demonstrated by the user. In the design of a workplace, attempts to use immersion should be made only when applicable, and real-world artifacts should be used otherwise.

The system has two 'alternative realities' for the users to explore:

- the 'real world', where they have the objects they are used to working with in GIS rooms
- the virtual world, where the user can immerse him- or herself for visualization purposes

The two environments are connected, and every action in the real world has an effect in the virtual one. Both worlds exist simultaneously and the HMD is a bridge between them. Each virtual tool has a physical counterpart and the changes that occur in the virtual world can also be observed from outside, making the immersion in the real and virtual worlds coexist, because sensorial information about the 'other' world remains accessible.

The overall system components are: the computer systems; a digitizing tablet, with the associated tablet mouse or pen; and a head-mounted display, with a Polhemus tracker attached. These components have different functions and act in the real and in the virtual environment. Any real object added to the system should be tracked and represented in the virtual space with an object chosen by the user.

The central objects of interaction are maps, both real and virtual, with a previously established correspondence activated by a calibration procedure. Any new object introduced in the system will always be related to the active map. The real-virtual correspondence is extended to the working space representation, i.e. the virtual space is a virtual GIS room with objects and operations similar to those that occur in the real space.

13.5 VIRTUAL ROOM DESIGN

The virtual room was modeled to imitate a real GIS room. The room is made up of a workbench, where the digital terrain models (DTM) are draped, and several layers of information (maps), represented as portraits on the wall. The users have direct access to the information in the portraits, as they hear a 3-D sound with a description of a layer when they select it. A blinking feedback is added for visual confirmation.

The visualization background is the terrain floating over the table. This terrain is scaled to fit the table, and is divided into a grid of regular cells, with the size depending on the spatial scale chosen by the user.

Any terrain size is intended to be placed floating over the table. As the data involved could have gigantic proportions, level-of-detail representation and visualization methods were designed. The intention is to give to the computer only the data that it needs to generate each frame.

When the user navigates on the terrain, the detail levels of the different areas vary. Hence, a quad tree was the representation chosen for the terrain layers. To use even fewer triangles, an irregular triangulation (TIN) is constructed for each quad tree leaf node. The TIN was designed to represent the elevation of the area with a high degree of cartographic coherence (Figure 13.1).

Figure 13.1 The TIN created to represent the digital terrain model. The textured terrain is blended with a wireframe version to highlight the TIN structure.

In real time, a level-of-detail management technique uses the quad trees previously stored in a database. Essentially, it expands the selected layers' quad trees in variously sized branches to meet the optimal level of detail for an area. It also assures smooth transitions between the different TINs.

Layer selection and visualization has very simple rules of interaction. The user selects the active layer by clicking in the appropriate tablet location, and the corresponding layer becomes the active layer: the following actions will all have that layer as input. After the selection, subsequent clicks on tablet mouse button 1 cause the drop of the layer only in the sub-terrain closer to the pointing device in virtual space. A click on tablet button 2 causes the dropping of the complete selected layer over the terrain. In a sense, this kind of interaction is like painting on a DTM, being used as a way of visualizing several layers simultaneously (see Figure 13.2).

A Polhemus is used on the head to track the user's position and orientation, providing a very natural method of navigation in the virtual environment. There is no need to use additional navigation devices, because the real table is in the range of the tracker. The digitizing tablet works as a world in miniature (Pausch and Burnette, 1995; Stoakley *et al.*, 1995), in which each action has a repercussion in the virtual world.

The tools are virtual objects that represent as clearly as possible the mental models that common GIS users have of a particular abstraction. For instance, fire models can be represented with a match object, and wind models with a fan. If the users intend to run one of these models, they just have to click on the appropriate table location, which makes that tool active, clicking again over the terrain to activate the model. The model will stop running after a click on tablet button 2.

The visual feedback of the tablet mouse depends on the active selection mode at that moment. The user will see a hand when in selection mode and a brush when in painting mode. These modes change automatically:

- if the tablet mouse pointer is over a poster, then it changes to a hand
- if the tablet mouse pointer is over the DTM, it changes to a brush

The tablet icon also changes according to the tool selected.

For performance reasons, the DTM is generalized (scaled) to be dropped over the table. If one intends to explore it in greater detail, a portion of the terrain can be selected and then be virtually transported, passing through a portal that connects the virtual GIS room to it. In this new universe, the user flies over the terrain in autopilot mode, with his or her head defining the viewpoint orientation.

13.6 APPLICATION

To illustrate the above ideas, an existing forest fires simulation application was interfaced and simple protocols to exchange information were designed.

The outputs of the model are seen as updates of the digital terrain model (DTM) lying over the table. There is a correspondence between cell sizes in the virtual environment and the simulation environment that is established when the model starts running. As new simulation steps are computed, the information about the cells that have changed is sent to the VE, reporting the cell location (x,y) and the new cell value (see Figure 13.2).

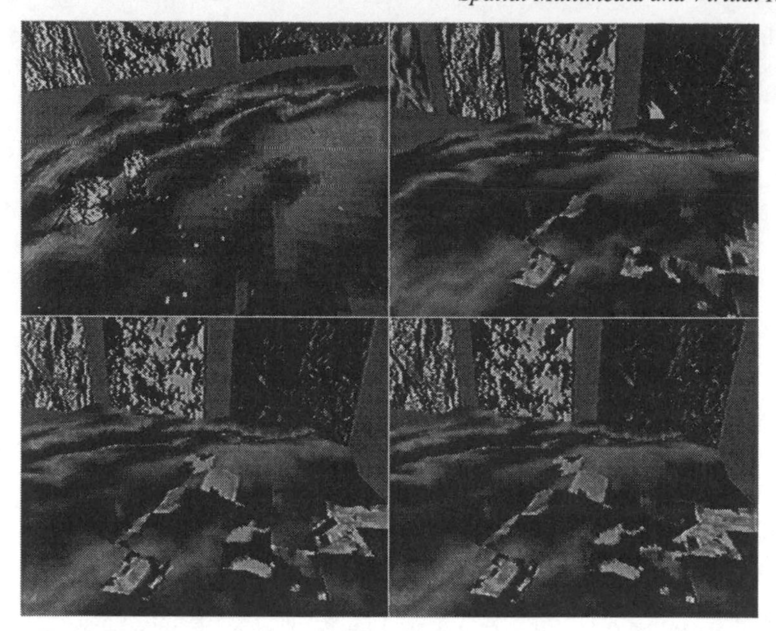

Figure 13.2 Several snapshots that describe the interaction in the virtual GIS room. From left to right, top to bottom: the forest fires visualization; layer selection; painting on the terrain (two bottom images).

13.7 IMPLEMENTATION ISSUES

This project was developed using a PC with a 60 MHz Pentium™ processor, and works with 16 MB of RAM and a 1 GB hard disk. An SPEA FIRE™ accelerator graphics board provides the images for the Virtual Research EyeGen3™ Head Mounted Display (HMD) through the RGB to NTSC Encoders. The tracking system is the Polhemus 3Space ISOTRACK™. The software used was WorldToolkit™ version 2.0, from Sense8.

13.8 CONCLUSIONS

The use of a virtual environment system to visualize and interact with spatial information and associated simulation models allows the user to explore information, thus re-creating their usual interaction in a real GIS room. The prototype developed was used by several users and their comments annotated and taken into consideration in minor modifications to the interface. The wall posters and their visualization in the DTM seemed to be easy to understand, without any previous knowledge of the concepts involved.

13.9 FUTURE DEVELOPMENTS

When this project started, the GIS systems on the market were not object-oriented, which made this approach more difficult and only feasible with the available GIS source code. The picture has considerably, now major GIS vendors are selling ActiveX (the substitute for OLE) objects that implement their products' methods. It is now possible to develop new versions of this project starting from a completely different level.

ACKNOWLEDGMENTS

This work was partially supported by PRAXIS XXI under research contract Praxis/3/3.2/AMB/04/94. The work of Nelson Neves is funded by Junta Nacional de Investigação Científica e Tecnológica (JNICT) through a PhD fellowship. The PRAXIS grant 3/3.2/AMB/04/94 supports the work of Pedro Gonçalves. The authors also wish to acknowledge the advice and corrections to the final version given by Professor António Câmara.

REFERENCES

BROCKSCHMIDT, K., 1994, OLE integration technologies, *Dr. Dobb's Special Report*, 225 (Winter), 42-9.

BURDEA, G. and COIFFET, P., 1994, *Virtual Reality Technology*, New York: John Wiley.

BURROUGH, P. A., 1986, *Principles of Geographic Information Systems for Land Resources Assessment*, Monographs on Soil and Resources Surveys, No. 12, Oxford: Oxford University Press.

CAMPAGNONI, F. R., 1994, IBM's System Object model, *Dr. Dobb's Special Report*, 225 (Winter), 24–8.

CHANCE, A., NEWELL, R. and THERIAULT, D., 1990, An object-oriented GIS: issues and solutions, in *Proceedings of the European Conference on Geographical Information Systems (EGIS) Annual Conference*, Utrecht, Netherlands, pp. 179–88.

CRUZ-NEIRA, C., SANDIN, D. J. and DEFANTI, T. A., 1993, Surround screen projection-based virtual reality: the design and implementation of the CAVE, in *Proceedings of ACM SIGGRAPH '93*, Anaheim, CA, ACM Press, pp. 135–42.

DIOTEN, R. and KOOY, J., 1995, Dynamic visualization of spatial data using virtual reality techniques, in *Proceedings of the Joint European Conference on Geographical Information*, The Hague, Netherlands, pp. 145–50.

FAIRCHILD, K. M., 1993, Information management using virtual-reality based visualizations, in WEXELBLAT, A. (Ed.), *Virtual Reality Applications and Explorations*, Cambridge, MA: Academic Press.

FISHER, M. M. and NIJKAMP, P., 1992, Editorial, *Annals of Regional Science*, **26**(1).

GOODCHILD, M. A., 1987, Spatial analytical perspective on geographical information systems, *International Journal of Geographical Information Systems*, **1**, 327–34.

HERRING, J., 1991, TIGRIS: a data model for an object-oriented geographic information system, *Computers and Geosciences*, **18**, 443–52.

JACOBSON, R., 1994, Virtual worlds capture spatial reality, *GIS WORLD*, December, 36–9.

MATTOS, N. M., MEYER-WEGENER, K. and MITSCHANG, B., 1993, Grand tour of concepts for object orientation from a database point of view, *Data and Knowledge Engineering*, **9**, 321–52.

MCGREEVY, M. W., 1993, Virtual reality and planetary exploration, in WEXELBLAT, A. (ed.), *Virtual Reality Applications and Explorations*, Cambridge, MA: Academic Press.

NCGIA, 1989, The Research Plan of the National Center for Geographic Information and Analysis, *International Journal of Geographical Information Systems*, **3**, 117–36.

NEWELL, R., 1992, Practical experience of using object-oriented programming to implement a GIS, in *Proceedings of the GIS/LIS Annual Conference*, Bethesda: ASPRS and ACSM, pp. 624–9.

OPENSHAW, S., 1987, An automated geographical analysis system, *Environment and Planning A*, **19**, 431–6.

PAUSCH, R. and BURNETTE, T., 1995, Navigation and locomotion in virtual worlds via flight into hand-held miniatures, in *Proceedings of ACM SIGGRAPH '95*, Los Angeles, CA, ACM Press, pp. 399–400.

ROBERTSON, P. K. and ABEL, D. J., 1993, Graphics and environmental decision making, *IEEE Computer Graphics and Applications*, March, pp. 25–7.

SCHEE, L. H. and JENSE, G. L., 1995, Interacting with geographic information in a virtual environment, in *Proceedings of the Joint European Conference on Geographical Information*, The Hague, Netherlands, pp. 151–6.

STOAKLEY, R., CONWAY, M. and PAUSCH, R., 1995, Virtual reality on a WIM: interactive worlds in miniature, in *Proceedings of ACM CHI 95*, Denver, CO, ACM Press, pp. 265–72.

STONEBRAKER, M., 1994, Sequoia 2000: a reflection on the first three years, *IEEE Computational Science and Engineering*, Winter, pp. 63–72.

TAYLOR, R. M., ROBINETT, W., CHI, V. L., BROOKS, F. P. JR., WRIGHT, W. V., WILLIAMS, R. S. and SNYDER, E. J., 1993, The nanomanipulator: a virtual-reality interface for a scanning tunneling microscope, in *Proceedings of ACM SIGGRAPH '93*, Anaheim, CA, ACM Press, pp. 127–34.

VALDÉS, R., 1994a, Introducing interoperable objects, *Dr. Dobb's Special Report*, 225 (Winter), 4–6.

VALDÉS, R., 1994b, Implementing interoperable objects, *Dr. Dobb's Special Report*, 225 (Winter), 62–7.

VAN OOSTEROM, P. and VAN DEN BOS, J., 1989, An object-oriented approach to the design of geographical information systems, *Computer and Graphics*, **13**, 409–18.

WILLIAMS, S. and KINDEL, C., 1994, The component object model, *Dr. Dobb's Special Report*, 225 (Winter), pp. 14–22.

WORBOYS, M. F., 1994, Object-oriented approaches to geo-referenced information, *International Journal of Geographical Information Systems*, **4**, 385–99.

Index

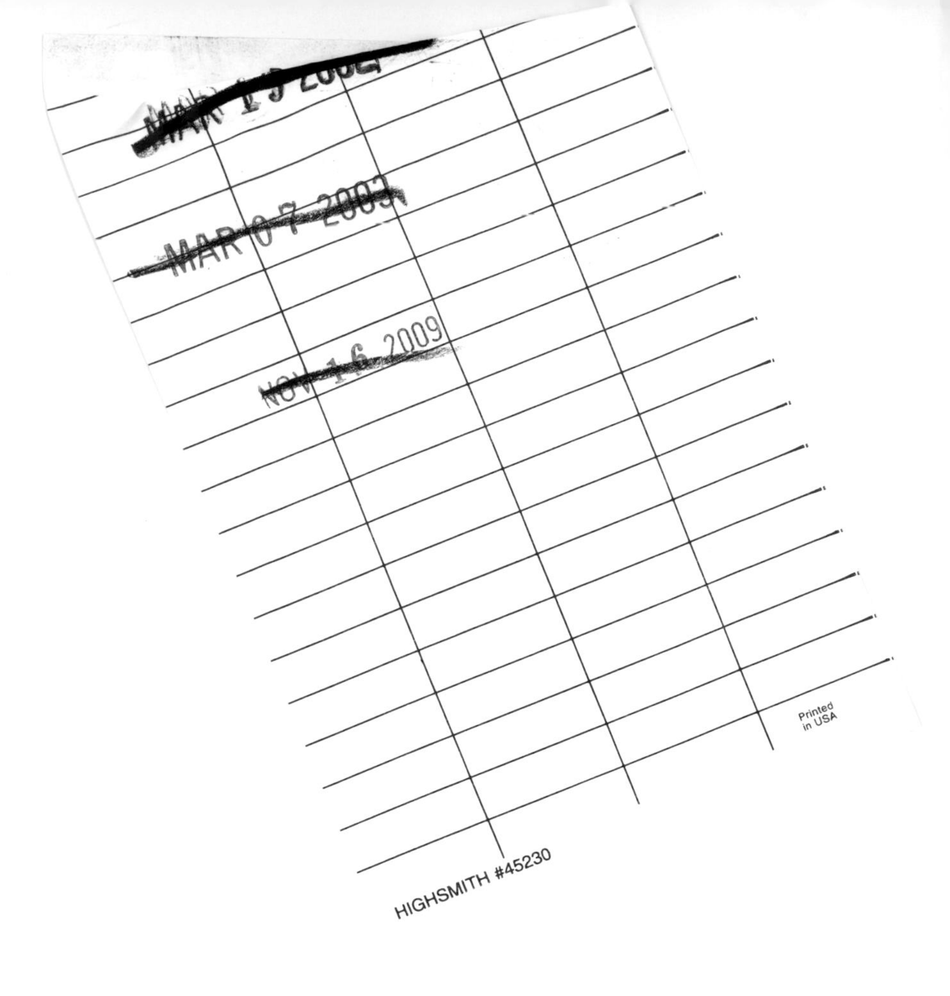

MAR 19 2002
MAR 07 2003
NOV 16 2009
Printed in USA
HIGHSMITH #45230